Lab Manual
to Accompany
Aquaculture Science

Third Edition

Delmar Cengage Learning
is proud to support
FFA activities

Join us on the web at

agriculture.delmar.cengage.com

Lab Manual
to Accompany
Aquaculture Science

Third Edition

Rick Parker

DELMAR
CENGAGE Learning™

Australia • Brazil • Japan • Korea • Mexico • Singapore • Spain • United Kingdom • United States

DELMAR
CENGAGE Learning™

Lab Manual to Accompany, Aquaculture Science, Third Edition
Rick Parker

Vice President, Career and Professional Editorial: Dave Garza

Director of Learning Solutions: Matthew Kane

Acquisitions Editor: Sherry Dickinson

Managing Editor: Marah Bellegarde

Senior Product Manager: Christina Gifford

Editorial Assistant: Scott Royael

Vice President, Marketing: Jennifer Baker

Marketing Director: Debbie Yarnell

Marketing Manager: Erin Brennan

Marketing Coordinator: Erin DeAngelo

Production Director: Carolyn Miller

Production Manager: Andrew Crouth

Senior Content Project Manager: Katie Wachtl

Senior Art Director: Dave Arsenault

For product information and technology assistance, contact us at
Cengage Learning Customer & Sales Support, 1-800-354-9706

For permission to use material from this text or product,
submit all requests online at **www.cengage.com/permissions**
Further permissions questions can be emailed to
permissionrequest@cengage.com

Library of Congress Control Number: 2011920462

ISBN-13: 978-1-4354-8810-6

ISBN-10: 1-4354-8810-5

Delmar
5 Maxwell Drive,
Clifton Park, NY 12065-2919
USA

Cengage Learning is a leading provider of customized learning solutions with office locations around the globe, including Singapore, the United Kingdom, Australia, Mexico, Brazil, and Japan. Locate your local office at: **international.cengage.com/region**

Cengage Learning products are represented in Canada by Nelson Education, Ltd.

To learn more about Delmar, visit **www.cengage.com/delmar**

Purchase any of our products at your local college store or at our preferred online store **www.cengagebrain.com**

Notice to the Reader
Publisher does not warrant or guarantee any of the products described herein or perform any independent analysis in connection with any of the product information contained herein. Publisher does not assume, and expressly disclaims, any obligation to obtain and include information other than that provided to it by the manufacturer. The reader is expressly warned to consider and adopt all safety precautions that might be indicated by the activities described herein and to avoid all potential hazards. By following the instructions contained herein, the reader willingly assumes all risks in connection with such instructions. The publisher makes no representations or warranties of any kind, including but not limited to, the warranties of fitness for particular purpose or merchantability, nor are any such representations implied with respect to the material set forth herein, and the publisher takes no responsibility with respect to such material. The publisher shall not be liable for any special, consequential, or exemplary damages resulting, in whole or part, from the readers' use of, or reliance upon, this material.

Printed in the United States of America
1 2 3 4 5 6 7 15 14 13 11

TABLE OF CONTENTS

INTRODUCTION

The Chinese understood learning styles many hundreds of years ago. An old Chinese proverb says, "I hear; I forget. I see; I remember. I do; I understand." Laboratory manuals are written to create the seeing to remember and the doing to understand. This brings real knowledge.

Two new laboratory exercises were added to this third edition. The 20 laboratory exercises in this manual were written to enhance learning of the material in the textbook *Aquaculture Science, Third Edition.* Labs in this manual are arranged in order of their correlation with the chapters in Aquaculture Science, Third Edition. Additionally, a new section in each lab suggests with which chapter in the text the lab can be used. Still, not all instructors will use every lab or all parts of every lab. Each instructor will need to read through the laboratory manual and decide what to use and when to use it.

Some of the labs may require a field trip or time out of the classroom. For example, Lab 2 requires going to a pond or stream and collecting plankton with a net, and Lab 8 requires the collection of frog legs. Lab 13 requires the students to use a Secchi disk after they construct it. Finally, Lab 3 requires students to interview twenty people—preferably not other students. Labs 15 and 16 assume that students have access to established aquariums and to a recirculating system.

Before using this manual in the classroom, the instructor should plan out the entire year or semester. Some of the labs can be completed within one lab period, but others will require two or more lab periods. Also, some labs use materials that need to be purchased ahead of time and need to be budgeted for. Other labs require minimal materials.

Knowledge gained from any of the laboratories in this manual will be increased if the instructor requires some sort of a written report. Writing a report clarifies and reinforces the concepts in the minds of the students. Of course, if instructors require written reports of the laboratories, they must read them and provide the students with feedback. Some will call this "writing across the curriculum." It is not a bad idea. For those who wish to try this, here is a simple outline for each report—

I. Introduction. (This states the purpose of the experiment and gives some brief background information.)
II. Materials and Methods. (This is a summary of the materials and general procedures used.)
III. Results and Discussion. (This contains the actual data obtained in the laboratory with the student's interpretation of the data and the conclusions reached. Any possible sources of error should also be discussed.)
IV. Summary and Application. (This section summarizes what was learned and how it is applied to aquaculture.)

Each lab exercise includes Introduction, Background, Materials, Procedures, and Analysis sections. These sections will be helpful for the student writing reports. But encourage the students to use their own words and not just copy out of the manual.

   The instructor may want to modify some of these labs so that they are more relevant to local needs and circumstances. Or, some instructors may want to include lab activities of their own creation. They should do so. Also, some of the exercises may stimulate additional, larger activities, or just the need to explore further. By all means, allow natural curiosity to run its course as long as it relates to the learning of aquaculture.

   Finally, the lab manual contains a short Appendix that provides the names and addresses for suppliers of equipment and specimens, laboratory rules, recipes for some of the chemical solutions, conversions for English-to-metric measurements, some do's and don'ts of recirculating systems, and an extra lab that can be used by some students depending on their access to the supplies and time.

This lab manual is dedicated to
the stimulation of creativity and
the pursuit of curiosity.

LAB 1

Virtual Tour of D.C. Booth Historic National Fish Hatchery

INTRODUCTION

Learning about fish biology and how to culture fish and develop fish hatcheries has been critical to the progress of a successful aquaculture industry. The D.C. Booth Historic National Fish Hatchery (Figure 1-1) played an early role and now serves as a living fishery museum.

The purpose of this lab is to introduce a key component, through a virtual tour, in the aquaculture industry's development.

FIGURE 1-1 Website for the D.C. Booth Historic Hatchery.

CORRELATION

This lab can be used with Chapter 1 or 5 of *Aquaculture Science*, 3rd Edition.

BACKGROUND

Established in 1896, D.C. Booth Historic National Fish Hatchery and Archives, formerly Spearfish National Fish Hatchery, is one of the oldest operating hatcheries in the country dedicated to fish culture and resource management. The hatchery was constructed to propagate, stock, and establish trout populations in the Black Hills of South Dakota and Wyoming. After a very successful fish production history, the hatchery ceased operations in the mid-1980s and reopened with a new mission and partnerships to help preserve the U.S. Fish & Wildlife Service's historic and cultural heritage.

Today, D.C. Booth Historic National Fish Hatchery still rears trout for the Black Hills through a cooperative effort with the State. The hatchery also serves to protect and preserve fishery records and artifacts for educational, research, and historic purposes, and to provide interpretive and educational programs to the public.

MATERIALS

➤ Internet access

➤ Word processor

PROCEDURES

1. Go to the website for the U.S. Fish & Wildlife Service, D.C. Booth Historic National Fish Hatchery and Archives: http://www.fws.gov/dcbooth/index.htm.

2. Read these sections on the website: History, Museum, Partnerships and Hatchery Tour (virtual).

3. Answer the questions in the "Analysis" section.

ANALYSIS

1. Where is the D.C. Booth Historic National Fish Hatchery located?

2. Who was D.C. Booth?

3. What was the original purpose of the D.C. Booth Hatchery?

4. What can be found in the museum and what connection does the name Von Bayer have to fish culture?

5. Based on the virtual tour of the D.C. Booth Hatchery, name eight sites you would visit at the hatchery.

6. What is the "Yellowstone Boat?"

7. Describe the railroad fish car and its use.

8. What partnerships are involved with the D.C. Booth Hatchery?

9. How does a person get into the Fish Culture Hall of Fame?

10. Provide the name and a brief description of one member of the National Fish Culture Hall of Fame (http://www.fishculturesection.org/Hall_of_fame/halloffame.html).

LAB 2

Use of Classification Keys for Fish and Some Microorganisms

INTRODUCTION

Being able to identify organisms is important to an aquaculturalist. Classification is a method of separating a large group of closely related organisms into smaller groups. A classification key makes identification easy by listing specific characteristics such as structure, form, and behavior in such a way that the identity of an organism can be deduced. Perhaps the most difficult part of using a key for identification is understanding the terms used to describe the organisms.

The purpose of this lab is to introduce the use of a key to identify some pictures of fish. Then, the use of another key is introduced to identify microorganisms collected from pond water.

CORRELATION

This lab can be used with Chapter 2 of *Aquaculture Science,* 3rd Edition.

BACKGROUND

Scientific classification is based on structural similarity. Biologists classify all living things into groups that show structural similarity. To do this, biologists start with large groups and continue dividing the groups until they arrive at a single kind or species of plant or animal. Carolus Linnaeus, an eighteenth-century Swedish botanist, devised a sound system for classifying and naming organisms. Linnaeus discarded common names and used scientific names made up of Latin words. In this system, an organism's name consists of two or more parts. The first part refers to the genus and begins with a capital letter. The second part refers to the species and usually begins with a small letter. A genus is a group of closely related species. A species is an individual or distinct kind of living thing like all others of its kind in form. A species is capable of producing more of its kind. When organisms of a single species vary slightly but not enough to be considered a separate species, they are called *varieties*. This adds a third name to the scientific name. Scientific names are written in italics, but if this is not possible, they are underlined.

Of all the animals with a backbone, fish are perhaps the most difficult to identify. Worldwide, about 30,000 species of fish exist. Some of these fish are widely distributed and may be given a common name useful only in one location. Common names are confusing and misleading. Classification keys that eventually identify a fish down to its scientific name are the most useful.

A classification key is used to identify common microorganisms found in ponds. In aquaculture, it is often important to know what microorganisms inhabit the same water as farmed species because these may be beneficial or harmful. A classification key and drawings verify the correct identification.

When organisms are found that are not included in a key, the biologist documents this with a drawing (photograph) and a description. Then, the biologist determines where the new organism fits in the classification key.

MATERIALS

These materials are necessary for the second part of the lab, when samples of pond water are collected so that the microscopic organisms can be identified.

- Dropper
- Pond water
- Microscope slide
- Coverslip
- Compound microscope
- Pencil
- Lab notebook
- Methyl cellulose
- Plankton net
- Quart jar

PROCEDURES

In the first part of this lab, you classify pictures of some fish. The key used classifies the fish to a common name. Although this is not ideal, it provides practice in using a classification key.

The second part of this lab requires the student to collect samples of pond water and then use a classification key to identify some of the plankton observed through a microscope. Classifying living organisms is considerably more difficult than classifying pictures.

Classification of Fish

Using the classification key in Table 2-1, name the fish in Figure 2-1.

1. Examine one of the drawings of a fish in Figure 2-1.

2. Read both statements listed under number 1 in Table 2-1. One of these statements should fit the fish you are trying to identify.

3. Refer to the number after the statement that fits the fish you selected. Go to this number in the key.

4. Again, read both statements and select the statement that best describes the fish you are trying to identify.

5. Continue through the key in Table 2-1 until you are not directed to "Go to" another line number in the table and you have the name of a fish instead of a "Go to." At this point the fish is named. Write the name of the fish on the line below the fish in Figure 2-1.

6. Select another fish from Figure 2-1 and repeat the process until all of the fish are named.

TABLE 2-1 CLASSIFICATION KEY TO CERTAIN FISH

1a. 1b.	Body more or less covered with scales Scales lacking or too small to be seen	Go to 2 Go to 12
2a. 2b.	Dorsal fin single Dorsal fins two or more, joined or separated	Go to 3 Go to 6
3a. 3b.	Body more than four times as long as broad (top to bottom); front edge of dorsal fin far back on body; mouth large, hinge back of eye Body less than four times as long as broad; front edge of dorsal fin about midway between head and tail; mouth not large, hinge in front of eye	Go to 4 Go to 5
4a. 4b.	Dark lines forming netted design on body; fins not spotted Body covered with yellow spots; fins spotted	Pickerel Northern pike
5a. 5b.	Mouth turned downward; barbels absent; dorsal fin not elongated Mouth not turned downward; barbels present; dorsal fin elongated	White sucker Carp
6a. 6b.	Two dorsal fins separated, the anterior spiny and the posterior soft Two dorsal fins united, forming an anterior spiny portion and a posterior soft portion.	Go to 7 Go to 8
7a. 7b.	Top of head concave, forming a hump in front of dorsal fin; dark vertical bars on body Top of head not concave, body sloping to dorsal fin and not forming a hump; dark blotches on body	Yellow perch Walleye
8a. 8b.	Body more than three times as long as broad Body less than three times as long as broad	Go to 9 Go to 10
9a. 9b.	Hinge of jaws behind the eye; notch between spiny and soft dorsal fin deep and nearly separating into two fins Hinge of jaws below the eye; notch between spiny and soft dorsal fin not nearly separating into two fins	Large-mouth black bass Small-mouth black bass
10a. 10b.	Mouth large, hinge below or behind eye Mouth small, hinge in front of eye	Go to 11 Bluegill
11a. 11b.	Five to seven spines in dorsal fin; dark spots forming broad vertical bars on sides Ten or more spines in dorsal fin; sides flecked with dark spots	White crappie Rock bass
12a. 12b.	Body much elongated and snakelike; dorsal, caudal, and anal fins continuous Body not elongated and snakelike; dorsal, caudal, and anal fins separate; adipose fin present	Eel Go to 13
13a. 13b.	Barbels growing from lips and top of head; head large and broad Barbels lacking; head not large and broad	Go to 14 Go to 16
14a. 14b.	Caudal fin deeply forked; head tapering Caudal fin rounded or slightly indented but not forked; head blunt	Go to 15 Bullhead catfish
15a. 15b.	Dorsal fin rounded at top; body silvery, speckled with black markings Dorsal fin long and pointed at top; body bluish-gray without speckles	Channel catfish Blue catfish
16a. 16b.	Caudal fin deeply forked; back not mottled and with few spots Caudal fin square or slightly indented; back mottled or spotted	Atlantic salmon Go to 17
17a. 17b.	Back and caudal fin spotted; broad horizontal band along sides Back mottled with dark lines; caudal fin not spotted; fins edged with white	Rainbow trout Brook trout

A. _____   B. _____   C. _____

D. _____   E. _____   F. _____

G. _____   H. _____   I. _____

FIGURE 2-1a Nine fish to classify using the classification key in Table 2-1.

Classification of Plankton

Using the classification key in Table 2-2, page 8, and the drawings in Figures 2-3 and 2-4, pages 10–12, collect a sample of pond water and identify the plankton found through a microscope.

1. To increase the number of plankton in the sample, pass a plankton net (commercially available) or nylon stocking through the pond and put the contents into the sample. Figure 2-2, page 9, illustrates a net that can be made from a nylon stocking. Cast the net into the pond and let it drift a few minutes before hauling it in. Another alternative is to wade into the water and pour water through the net for a few minutes. Finally, the net can be pulled through the water if a rowboat is available.

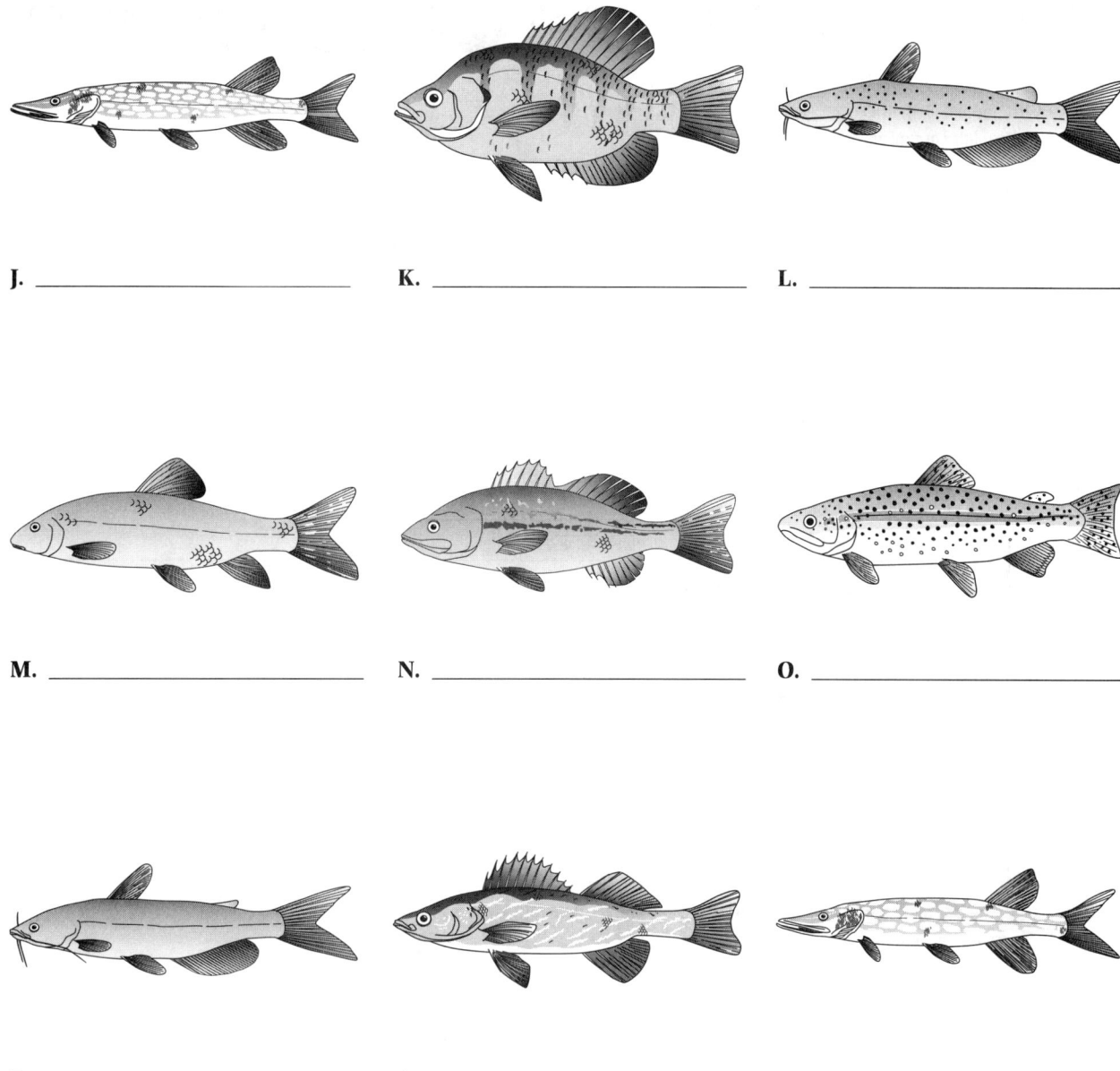

J. _____   K. _____   L. _____

M. _____   N. _____   O. _____

P. _____   Q. _____   R. _____

FIGURE 2-1b Nine fish to classify using the classification key in Table 2-1.

2. Collect the sample in a clean one-quart bottle that is one-half to two-thirds full of pond water. As the net is retrieved, quickly and carefully turn the contents into the jar.

3. Prepare microscope slides using a drop of pond water and a drop of methyl cellulose to slow the plankton movement. Add a cover slip and examine the sample at different levels of magnification.

TABLE 2-2 CLASSIFICATION KEY FOR POND WATER

1a.	Cells single or, if undergoing cell division, found in pairs	Go to 2
1b.	Cells numerous, multicellular; arranged in chains, filaments, or other multicellular organism	Go to 9
2a.	Cells or cultures of cells green in color	Go to 3
2b.	Cells or cultures of cells not green or a shade of green; generally yellow to brown in color	Go to 5
3a.	Cells motile by cilia	ciliated protozoan
3b.	Cells nonmotile by flagellum	Go to 4
4a.	Cells with bright green (e.g., grass green) chloroplast, larger (10 to 15 microns in diameter), oval or flattened shape	*Tetraselmis*
4b.	Cells a shade of green, generally small (2 to 5 microns in diameter), spherical or oval in shape	*Nannochloropsis*
5a.	Cells with radial symmetry	(centric diatoms) Go to 6
5b.	Cells not radially symmetrical, with somewhat bilateral symmetry (pennate diatoms)	Go to 7
6a.	Cells large, spherical, or oval, usually connected in chains of two or more cells, spines not present	*Melosira*
6b.	Cells smaller; square, rectangular, or oval; single or in short chains, spines originating at the corner of each cell	*Chaetoceros*
7a.	Cells tapering to a long, thin spine	*Nitzschia closterium*
7b.	Cell endings rounded or in a point, but not tapering to a long spine	Go to 8
8a.	Cells bilaterally symmetrical in all views	*Navicula* or other navicula-type diatom
8b.	Cells asymmetrical in some views	*Achnanthes*
9a.	Cells photosynthetic, arranged in a filament, variable in color	Go to 10
9b.	Cells non-photosynthetic, arranged into tissues and organs, generally clear or may be pigmented	Go to 12
10a.	Cells arranged as a branching filament, green in color	*Cladophora*
10b.	Cells arranged in a non-branching filament, brown in color	Go to 11
11a.	Cells generally in long chains of large, spherical cells, spines not present	*Melosira*
11b.	Cells in short or long chains or smaller, square, or rectangular cells possessing spines, which originate at the corner of each cell	*Chaetoceros*
12a.	Body with segmented body parts, appearing shrimplike, lacking cilia around the mouth	copepod
12b.	Body smooth, non-segmented, appearing saclike, with cilia around the mouth	rotifer

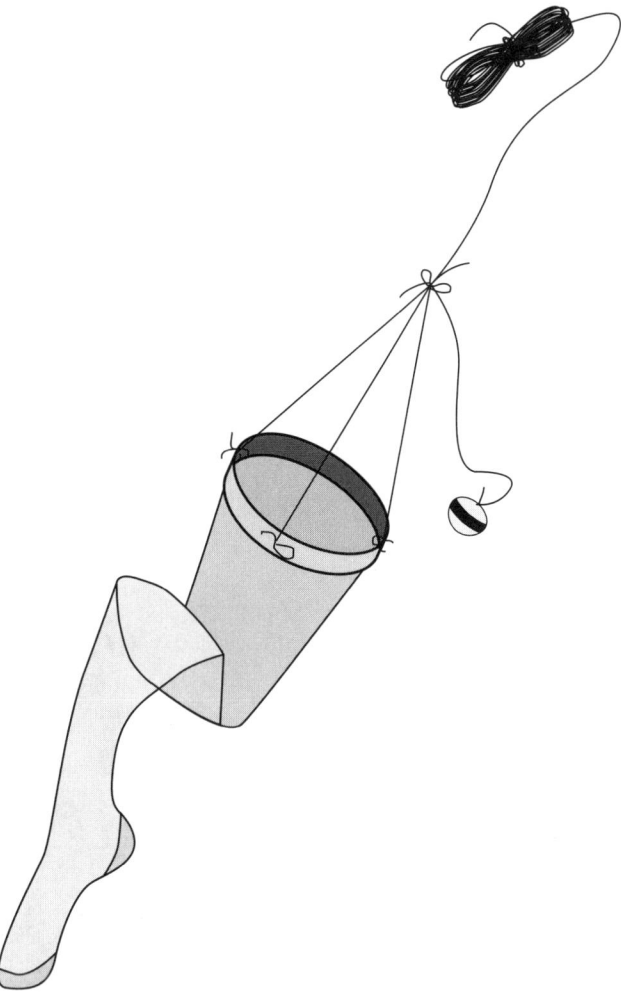

FIGURE 2-2 Example of a net that can be constructed to collect plankton.

4. Starting with low power (4X), locate the material. Move carefully to high power (10X) and focus. Examine the sample for the presence of plankton.

5. When one of a plankton is clearly in view, identify it, using the key in Table 2-2.

6. Read the first section of the key that contains two descriptions. The plankton you are attempting to identify fits either one or the other description. When you have decided which of the two descriptions best fits your specimen, go on to the next section and continue until you have identified your specimen.

7. Verify your choice with the illustrations provided in Figures 2-3 and 2-4a and 2-4b, and record the name in your lab notebook.

8. If there is a specimen that is not in the key, make a sketch in your lab notebook and identify it using other reference materials such as *Pond Life: Revised and Updated* (A Golden Guide from St. Martin's Press) by George K. Reid.

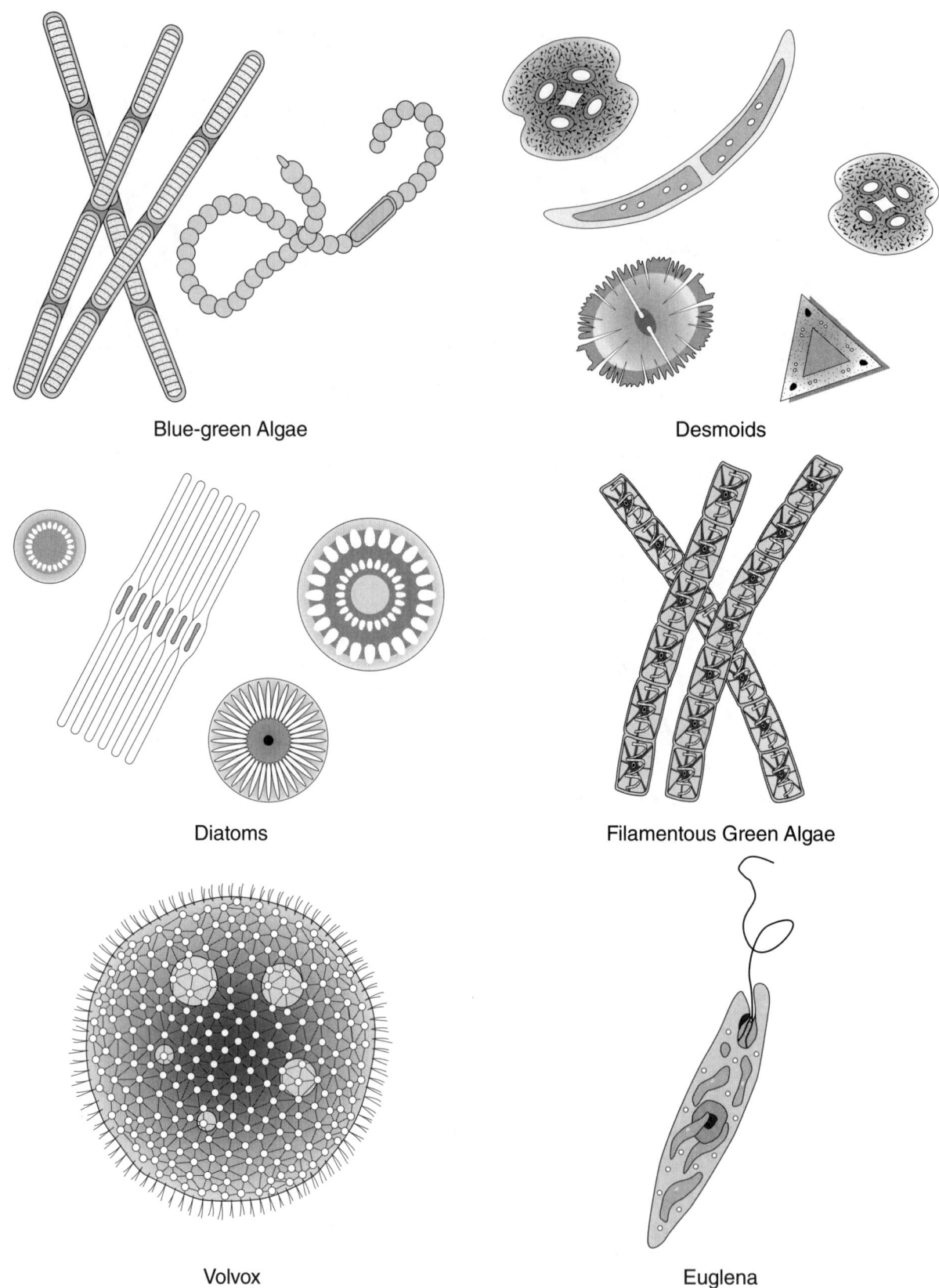

Blue-green Algae

Desmoids

Diatoms

Filamentous Green Algae

Volvox

Euglena

FIGURE 2-3 Examples of phytoplankton in pond water.

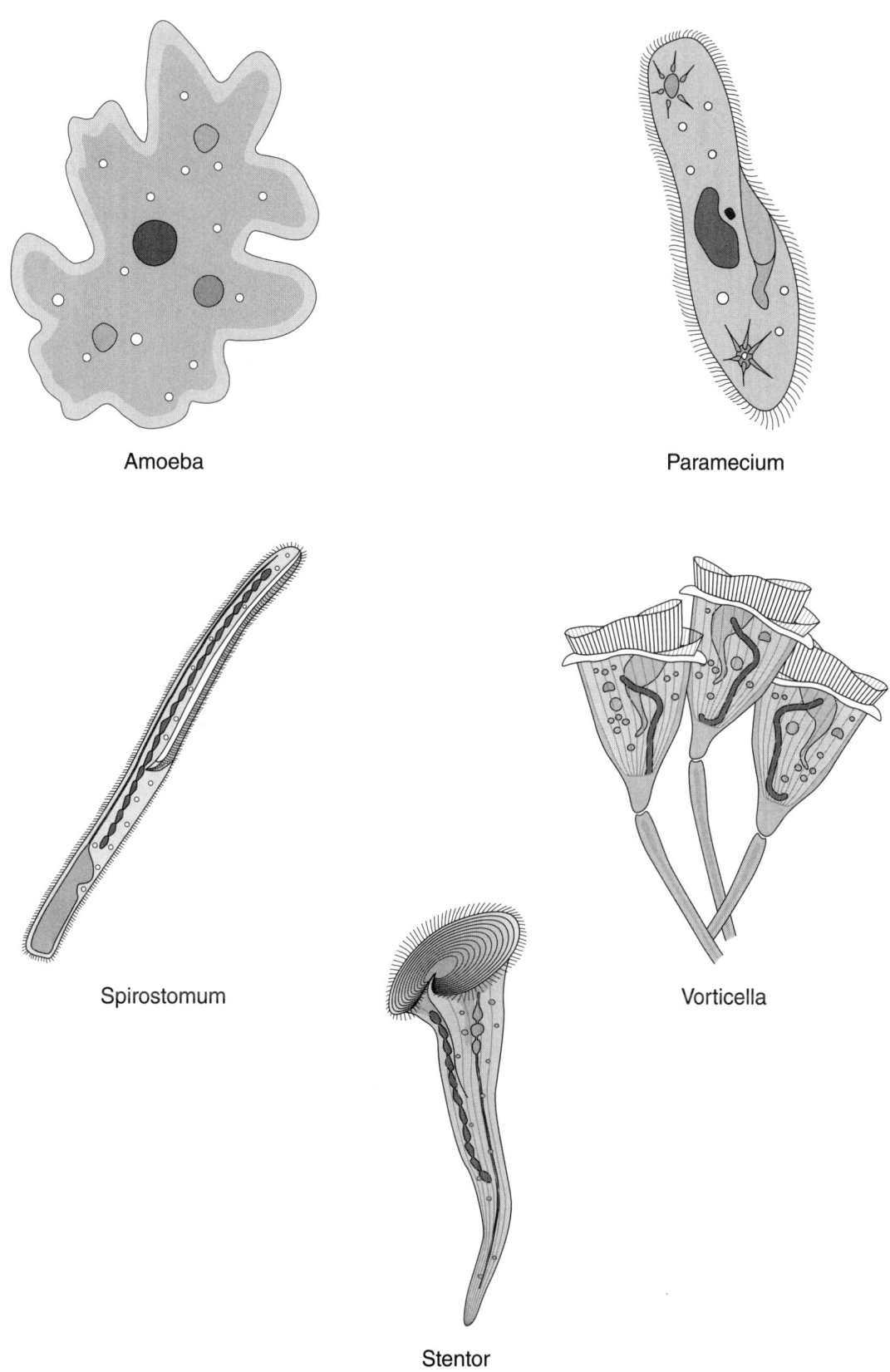

Amoeba

Paramecium

Spirostomum

Vorticella

Stentor

FIGURE 2-4a Examples of zooplankton in pond water.

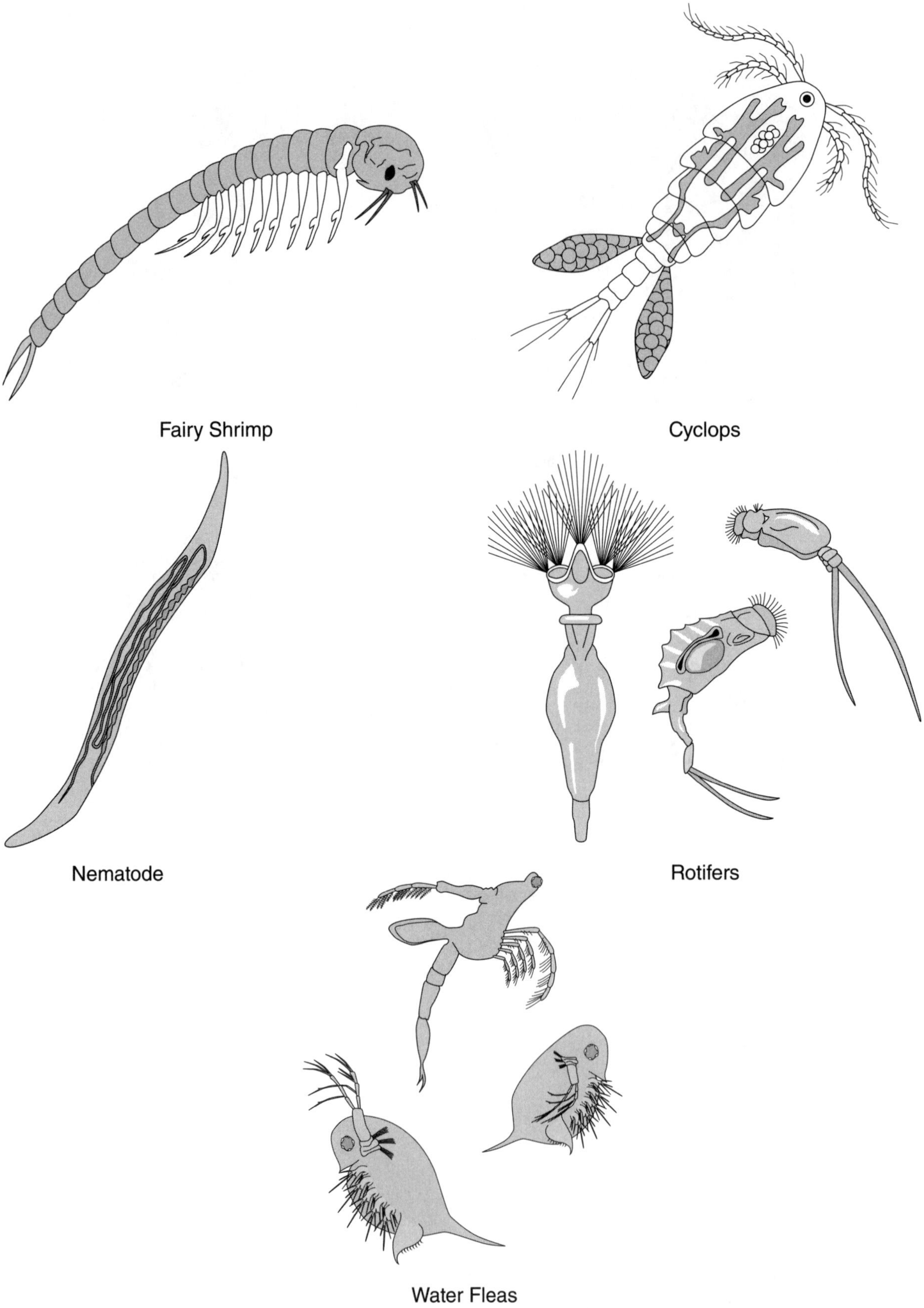

Fairy Shrimp

Cyclops

Nematode

Rotifers

Water Fleas

FIGURE 2-4b Examples of zooplankton in pond water.

ANALYSIS

1. What features of a fish are used in classification?

2. What features of plankton are used in classification?

3. Identify some of the problems you encountered trying to classify organisms.

4. Using the *Aquaculture Science, 3rd Edition*, textbook, complete Table 2-3.

TABLE 2-3 CORRELATION OF COMMON NAME WITH SCIENTIFIC NAME		
Figure 2-1 ID	Common Name	*Scientific Name*
A		
B		
C		
D		
E		
F		
G		
H		
I		
J		
K		
L		
M		
N		
O		
P		
Q		
R		

5. Describe how a biologist would begin constructing a classification key.

6. Why are classification keys and scientific names important to the science of aquaculture?

7. Identify some of the weaknesses of using common names.

LAB 3 *Market Survey*

INTRODUCTION

Often before a marketing plan is written or to refine a marketing plan, market surveys are conducted. These surveys provide an indication of the demand for a new product or the idea for a new product. Market surveys are conducted on a small sample of the population and the results are projected for the whole population. Some companies specialize in conducting very sophisticated marketing surveys.

The purpose of this lab is to conduct and analyze a simple market survey.

CORRELATION

This lab can be used with Chapter 3 of *Aquaculture Science*, 3rd Edition.

BACKGROUND

Marketing is the process of getting the product from the producer to the consumer. It is the final step in food production, but it should rate top priority in the mind of an aquaculturalist. Although the aquaculturalist may possess the skills and resources to grow a crop, his or her efforts are in vain without a place to sell the product. This means developing a marketing strategy or plan.

The first step in marketing is to understand the situation, the current production, and consumption of a product. Next, producers need to understand the marketing—its functions and strategies.

Depending on the operation, marketing plans can be a long document or an informal plan of several pages. Development of a plan allows the producer to analyze the opportunities and needs. With this done, the producer can focus on production and make better decisions. Market plans or strategies contain three key elements:

1. Determination of the present situation

2. Determination of market goals

3. Developed plans to reach the goals

MATERIALS

➤ Market Survey, Table 3-1

➤ Pencil

➤ Calculator

➤ List of potential people to interview

TABLE 3-1 AQUACULTURE MARKET SURVEY						
	Tally	**Total**			**Tally**	**Total**
1. What is your age range?			6. Which of the following are you likely to purchase?			
15–20			Trout			
21–30			Crawfish			
31–40			Catfish			
41–50			Shrimp			
Over 50			7. How much will you pay?			
2. What is your gender?			$3 to $5			
Male			$6 to $10			
Female			$11 to $20			
3. How many times per week do you eat fish or seafood?			$20 or more			
			8. Why do you purchase fish or seafood?			
None			Price			
One			Flavor			
Two			Variety			
Three or more			Health			
4. Where is it purchased?			9. In what form do you prefer to buy the product?			
Grocery store			Fresh			
Restaurant			Frozen			
Fish market			Canned			
Other			Other			
5. What method is used to prepare the fish or seafood?			10. How far will you drive to make a purchase?			
Broil			1 to 5 miles			
Bake			6 to 10 miles			
Fry			11 to 20 miles			
Steam			More than 20 miles			
Other						

PROCEDURES

1. Interview twenty people and ask the questions on the Market Survey.

2. Make tally marks in the Tally column to record responses during each interview.

3. After all 20 interviews, total the tally marks in the Total column.

4. Write a brief report summarizing your findings.

ANALYSIS

1. What problems did you encounter conducting your survey?

2. Using the averages from your survey, project how many people in your community are likely to buy trout, crawfish, catfish, or shrimp.

3. Based on your survey, who eats more fish or seafood—males or females, younger people or older people?

4. Where do most people in your community eat fish or seafood?

5. If you were trying to sell a new product, in what form would you sell it—fresh, frozen, or canned?

6. Most surveys are conducted on a random sample. Define a random sample.

LAB 4 — Anatomy of the Fish

INTRODUCTION

Many species of bony fishes exist in fresh, brackish, and salt waters. Their body forms range from the familiar perch, bass, and trout to shapes such as slender, snakelike eels and morays, deep-bodied ocean sunfish, and inflated boxfish. Exterior coverings of fish vary. For example, bony plates cover sturgeon, whereas scales cover the skin of many fish, such as trout and carp. Scales grow as the fish grows. A few species, such as the catfish, have skin without scales. The gills are in a common cavity in the pharynx, usually covered on each side by a platelike operculum. Fin number, placement, form, and structure differ between species, and these differences are useful in classifying fishes.

An understanding of the anatomy, the internal or external structure, of aquatic animals is essential for the successful aquaculturalist. Anatomy aids in distinguishing between the sexes and in spotting and diagnosing disease.

The purpose of this lab is to introduce the anatomy of a fish through an examination of the external features and dissection.

CORRELATION

This lab can be used with Chapter 4 of *Aquaculture Science,* 3rd Edition.

BACKGROUND

Physiology, or the function of aquatic animals, occurs in body systems. These systems in aquatic species are adapted to the water environment. Nine body systems are found in animals, including aquatic animals. These systems are:

1. Skeletal
2. Muscular
3. Digestive
4. Excretory
5. Respiratory
6. Circulatory

7. Nervous

8. Sensory

9. Reproductive

Skeletal System

The skeletal system is the rigid framework giving the body shape and protecting the organs. Muscles and organs attach to the skeleton. Fish possess an internal skeleton or endoskeleton.

Muscular System

The muscular system provides movement internally and externally. Muscles vary in strength and function. Muscles contract and relax to cause movement. Organisms require movement for such functions as obtaining food and oxygen and eliminating wastes.

Digestive System

The digestive system converts feed into a form that can be used by the body for maintenance, growth, and reproduction. It consists of all the parts of an organism involved in taking food into the body and preparing it for incorporation into the body. In its simplest form, the digestive system is a tube extending from the mouth to the anus with associated organs. In most species this includes the mouth, esophagus, stomach, intestines, anus, and other associated organs, such as the liver.

Digestive systems vary according to whether the animals are herbivores eating only plants, carnivores eating only animals, or omnivores eating plants and animals. Fish consuming algae and detrital matter have small stomachs and long intestines. Carnivorous fish possess large stomachs and short intestines.

Excretory System

Life processes produce waste products. The excretory system eliminates wastes from the body. Typically it consists of the gills, kidneys, urinary ducts, urinary bladder, and urinary opening. Kidneys filter the wastes from the blood. The urinary bladder holds the wastes until they are excreted through the urinary opening.

Respiratory System

The respiratory system takes in oxygen from the environment, delivering it to the tissues and cells of the body. The respiratory system also picks up carbon dioxide from the tissues and cells delivering it to the environment. Gills are the respiratory organs of fish. Water taken in is forced over the gills where oxygen is removed by diffusion into the blood. Carbon dioxide in the blood is higher than in the surrounding water. Carbon dioxide diffuses out of the blood through the gills. Cell membranes of gill cells are very thin and semipermeable, allowing gases to pass through. Figure 4-1 illustrates the anatomy of the gills and the process.

Circulatory System

The circulatory system distributes blood throughout the body. This system consists of a heart, veins, and arteries. Pumping action of the heart causes blood to flow through the arteries to the gills where it picks up oxygen and carries it to the rest of the body. Oxygen is necessary for all cells of the body. As the blood delivers oxygen to the cells, it picks up carbon dioxide, a waste product, which is carried in the blood back through the veins to the heart and gills. The gills release the carbon dioxide to the environment and pick up more oxygen. Blood pumped to the cells and tissues of the fish body also carry nutrients and waste products.

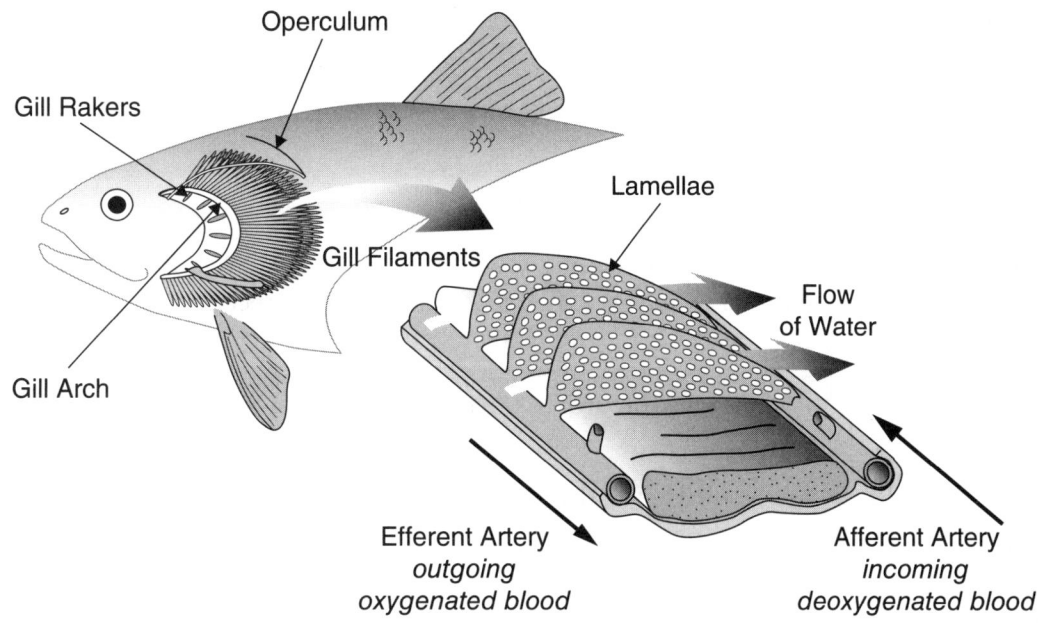

FIGURE 4-1 Anatomy of the gills and how they work.

Nervous System

The nervous system supplies the fish with information about its internal and external environment. This system conveys sensation impulses—electrical-chemical changes—between the brain or spinal cord and other parts of the body. It consists of the brain, spinal cord, many nerve fibers, and sensory receptors. It is a complex system. The sense organs or receptors receive stimuli and convey these by the nerve fibers to the brain or spinal cord where they are interpreted. The brain or spinal cord may send responses back through the nerve fibers.

Sensory System

The sensory system includes the five senses—sight, touch, taste, smell, and hearing. The sensory system relays information through the nervous system. Fish use eyes to find food and identify predators. Ear bones in the skull pick up water vibrations as sound. The sense of taste is important to the aquafarmer when selecting and preparing feed for fish. Some species have an enhanced sense of touch, through organs such as the barbels on catfish. Lateral lines in fish contain nerves that detect water vibrations and motion. The lateral line helps the fish maintain balance and position in the water.

Reproductive System

Sexual reproduction is the process of creating new organisms of the same species through the union of the male and female sex cells—sperm and eggs. Males and females exist in most species. Testes in the males produce sperm. Ovaries in the females produce eggs or ova. Testes and ovaries are located in the body cavity of the fish.

MATERIALS

➤ Fish—fresh, frozen, or preserved

➤ Dissecting tray

➤ Dissecting scissors

➤ Blunt probe

➤ Scalpel

➤ Forceps

PROCEDURES

This lab examines the external and internal anatomy of a fish. Before doing any dissection, begin by examining the external regions and structures. Then proceed with the dissection.

External Anatomy

Figure 4-2 shows some of the external structures of the fish. Use Figure 4-2 to help examine and identify structures and regions.

1. Before identifying any structures locate the following surfaces or regions:
   - Anterior
   - Posterior
   - Dorsal
   - Ventral
   - Lateral

2. Identify the following body regions:
   - Head (tip of snout to posterior edge of operculum)
   - Trunk (operculum to anus)
   - Tail (anus to end of caudal fin)

3. Identify the median fins:
   - Dorsal (usually two; anterior supported by spiny rays, posterior by soft rays)
   - Anal (one, ventral surface behind anus)
   - Caudal (one, end of tail, symmetrical dorsal and ventral lobes, soft rays only)

4. Identify the paired fins:
   - Pectoral (two, anterior part of trunk)
   - Pelvic or ventral (two, position on body varies according to species)

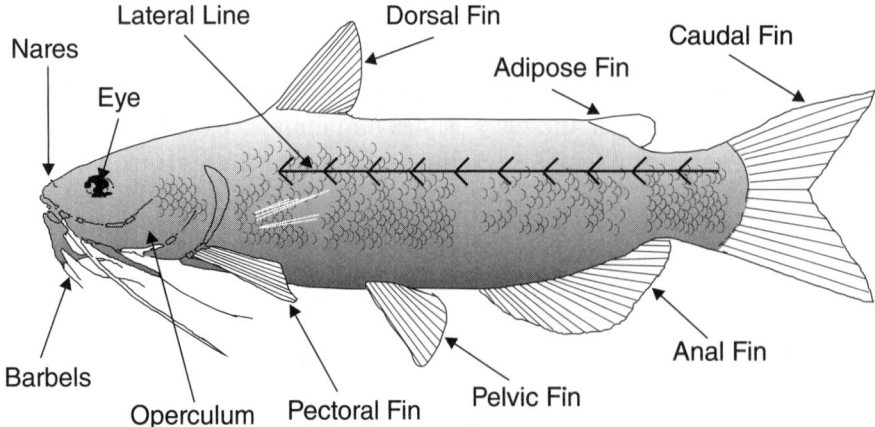

FIGURE 4-2 External structures of the fish.

5. On the head, identify the following:
   - Nostrils or nares (two, dorsal)
   - Eyes (two, lateral)
   - Mouth
   - Operculum (two, plate with bony support covering gill chamber)

6. On the trunk, identify:
   - Scales
   - Lateral line
   - Anus (mid-ventral near base of anal fin)
   - Urogenital opening (between anus and anal fin)

Fin Structure. Examine one of the paired fins, scraping off the skin to reveal the supporting rays. Determine the difference in structure between a spiny fin ray and a soft fin ray.

Scales. If your fish has scales, use a scalpel to remove one scale from the lateral line region, mount in water, and examine under low magnification on a compound microscope. Does the free edge have minute teeth? Can lines indicating growth (age) be seen?

Gills

Gills are in a common cavity at either side of the pharynx.

1. Cut off the operculum on the left side to find the four gills.
2. Probe through the mouth to learn the path of respiratory water flowing in over the gills.
3. Cut entirely across one gill, remove a small section, and examine under water to see the double row of gill filaments.
4. Notice that each gill is supported on a bony arch.
5. Find the slender gill rakers on the inner surface of each arch. They protect the delicate gills from food passing through the pharynx.

Muscles

Remove the skin and scales on the left side between the anal and caudal fins to expose the muscles beneath. The muscles are divided into shaped muscle segments (myomeres) extending from mid-dorsal to mid-ventral line. Each segment is separated from those adjacent by connective-tissue partitions (myosepta). A lengthwise partition beneath the lateral line separates each segment into dorsal and ventral parts.

Skeleton

The head is composed of about 40 bones. The skeleton surrounding the body organs forms a compact movable support for the gill mechanism. The pectoral fins join the pectoral girdle that attaches dorsally behind the head. The pelvic fins move on two flat bony plates lying in the ventral trunk muscles. The vertebrae are more delicate, proportionately, than those in most land vertebrates. The vertebral column ends in several stout flat bones supporting rays of the caudal fin. The dorsal and anal fin rays articulate on intramuscular bones that alternate with the long dorsal and ventral spines of the vertebrae. Paired, slender riblike bones attach to each trunk vertebra.

Internal Structure

1. Cut the body wall along the mid-ventral line from just in front of the anus to the lower jaw. Be careful not to injure organs within.

2. On the left side, cut dorsally at either end of this and remove the body wall to expose the body cavity and internal organs.

3. Use Figures 4-3 and 4-4 to help identify some of the following structures:

   - Peritoneum (a glistening membrane lining the coelom and covering organs)
   - Mesenteries (the double layer of peritoneum supporting organs mid-dorsally)
   - Heart (located in the pericardial sac, beneath gill region)
   - Stomach (a saclike organ varying in size, depending on food content)
   - Intestine (a slender tube extending from stomach to anus)
   - Pyloric caeca (slender, blind pouches attached anteriorly on intestine)
   - Liver (a large organ of three lobes located anteriorly in the body cavity)
   - Gall bladder (a small greenish structure, with bile duct to intestine)
   - Spleen (a spherical, small, dark reddish organ located between coils of intestine)
   - Swim bladder (a long, thin-walled sac, dorsal in body cavity and attached to pharynx in some fish)
   - Kidneys (slender, dark organs located along the dorsal wall, above swim bladder)
   - Urinary bladder (joined by the fine ureter from each kidney, discharging through urogenital opening)
   - Gonads (two testes in male and two ovaries, united, in female, discharging through ducts to urogenital opening)

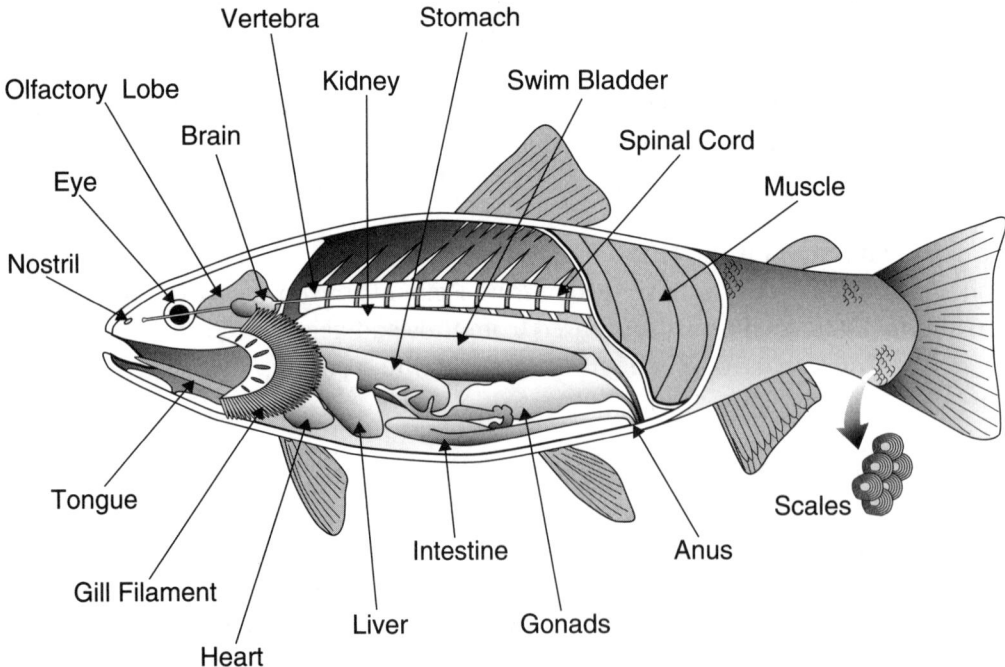

FIGURE 4-3 Internal anatomy of the fish.

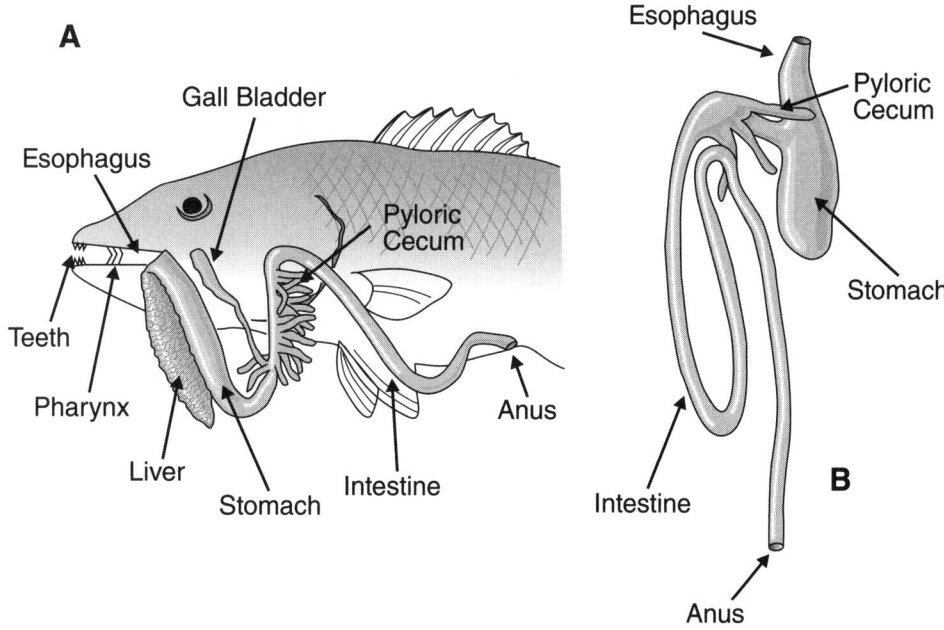

FIGURE 4-4 Anatomy of the digestive system of the fish.

ANALYSIS

1. Do the nostrils connect to the mouth cavity?

2. How are olfactory (smell) stimuli received?

3. Do the eyes have lids?

4. What sort of vision is possible?

5. Does the mouth open straight ahead, up, or down?

6. Is the mouth large or small compared to the size of the head?

7. Are the lips thin or thick?

8. Based on your examination of the mouth, where in the environment would you expect this fish to feed? Explain your reasoning.

9. Examine the teeth. How many teeth would you estimate?

10. Where in the mouth are the teeth located?

11. Describe the structure of the teeth. Are they narrow, sharp, dull, flat, or some other shape?

12. Based on your examination of the teeth, what would you expect this fish to eat? Explain your reasoning.

13. Describe the size and shape of the gill rakers.

14. What are the functions of the gills and gill rakers?

15. Is the digestive tract small, short, straight, coiled, or some other shape?

16. Measure and record the length of the entire digestive tract in centimeters.

17. Measure and record the length of the small intestine.

18. Determine what percentage of the digestive tract is small intestine.

19. Cut open the stomach. Remove the stomach contents, put them in a watch glass, and examine them under a magnifying glass. Can you identify what the fish has eaten recently?

20. Watch a live fish in an aquarium to learn the sequence of movements of the jaws and opercula in the passage of water for respiration.

LAB 5 *Culturing Shrimp*

INTRODUCTION

Brine shrimp are tiny crustaceans that live in saltwater. They are not the same kind that humans eat, but they are ideal for study, and they are often used to feed aquarium fish. Only a day or two is needed for eggs to hatch into young larvae (see Figure 5-1). Once hatched, the larvae grow into adult shrimp within two or three weeks.

The purpose of this lab is to observe the changes in appearance that occur as the shrimp grow and develop. This will be done by drawing pictures of the shrimp to provide a record of change.

CORRELATION

This lab can be used with Chapter 6 of *Aquaculture Science*, 3rd Edition.

BACKGROUND

Brine shrimp, *Artemia* species, have been used successfully for years as a food for aquarium fish. Brine shrimp eggs are noted for their viability. They are packaged and sold throughout the world. They arrive ready for hatching and feeding to small fish of all species. Stored in a cool dry place, brine shrimp eggs can last for at least ten years.

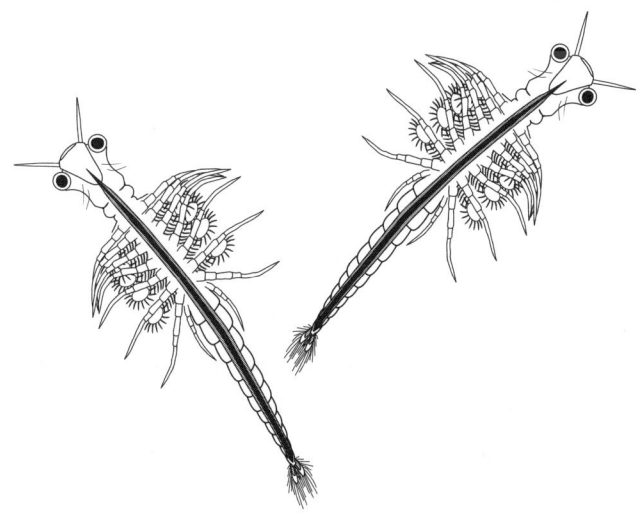

FIGURE 5-1 Brine shrimp naupalii a few hours after hatching.

Newly hatched brine shrimp are used as a first food for hard-to-feed species and small larval species, such as shrimp, walleye, and hybrid bass. They are easy to hatch and available at the time needed. They do not carry freshwater parasites or other organisms. If overfed to the fish, they do not cause rapid deterioration of the water quality.

MATERIALS

➤ Brine shrimp eggs (These are available in most aquarium shops, but do not buy them frozen.)

➤ Rock salt (un-iodized)

➤ Flat pan (pie pan), glass

➤ Cardboard cover

➤ Water

➤ Tablespoon

➤ Milk carton (empty, clean, one gal)

➤ Milk carton (empty, clean, one qt)

➤ Dropper

➤ Hand lens and dissection microscope

➤ Aquarium heater

➤ Air pump with air stone

Note: An aquarium tank may be used for the entire class.

PROCEDURES

To ensure hatching, maintain water temperature at 70° to 80°F with aquarium heater. Slight aeration with an air pump and air stone may also help.

1. Place several shrimp eggs under a lens or microscope and draw what you see in the space provided in Table 5-1.

2. Prepare the saltwater for the shrimp. Add 6 tablespoons of rock salt to 1 gal of water in the milk carton and wait until the salt dissolves. This will be the salt solution used to hatch the brine shrimp eggs.

3. Cover the bottom of a flat glass pan with saltwater. Add 1/6 tablespoon of the brine shrimp eggs to the saltwater in the pan. This is just enough to cover the tip of the spoon.

4. Put your name and the date on the cardboard cover.

5. Cover the pan with cardboard, leaving one end open about 1 in. **Note:** The larvae will be attracted to the light and will collect at the open end of the pan.

6. Each day, remove some animals from the pan and observe them with a hand lens or microscope. Draw them exactly as they appear to you in the space provided in Table 5-1.

TABLE 5-1 DRAWINGS OF THE DEVELOPMENT OF BRINE SHRIMP

Eggs, Date _____	Day 1, Date _____
Day 2, Date _____	Day 3, Date _____
Day 4, Date _____	Day 5, Date _____
Day 6, Date _____	Day 7, Date _____
Day 8, Date _____	Day 9, Date _____

7. After three or four days, use the dropper to temporarily transfer as many shrimp as possible to a quart carton filled with saltwater. Empty the pan and replace it with fresh salt water. Return the shrimp to the pan.

Note: Replace the cardboard cover on the pan. This helps to ensure continued growth of the shrimp that hatched.

ANALYSIS

1. Do your drawings show differences in appearance as the shrimp changes from egg to adult?

2. Why are the brine shrimp ideal for studying growth and development?

3. Explain what changes take place in the brine shrimp as it grows and develops.

4. Explain why brine shrimp are ideal subjects for studying growth and development.

5. Weigh 6 tablespoons of salt. Based on this weight, calculate the concentration of salt in 1 gal. of water in mg/l. How does this compare to the salinity of the oceans?

LAB 6   *Anatomy of a Crustacean*

INTRODUCTION

Crustaceans include shrimp, prawns, lobsters, crabs, and crawfish (crayfish). Most of the crustaceans considered for aquaculture are known as decapods—10 legs. The exoskeleton protects the soft body and gives support to the body. As a crustacean grows, the shell or exoskeleton is cast off in a process called *molting*. When crustaceans molt, they are known as *softshell* animals. No more than a day is usually required to regrow the shell. During molting, crustaceans are subject to attack by other aquatic animals, including their own species.

Through a process known as *regeneration*, crustaceans regrow limbs that have broken off.

In this laboratory, you will dissect a crawfish and learn about the form and function of its anatomy.

CORRELATION

This lab and/or Lab 7 can be used with Chapter 7 of *Aquaculture Science*, 3rd Edition.

BACKGROUND

The physiology, or function, of aquatic animals, occurs in body systems. In aquatic species, these systems are adapted to the water environment. Nine body systems are found in animals, including aquatic animals. These systems are:

1. Skeletal

2. Muscular

3. Digestive

4. Excretory

5. Respiratory

6. Circulatory

7. Nervous

8. Sensory

9. Reproductive

Skeletal System

The skeletal system is the rigid framework giving the body shape and protecting the organs. Tissues and organs attach to the skeleton. Crustaceans all possess an exoskeleton made of chitinous material. Chitin is a polysaccharide—a carbohydrate—of a hexose (sugar) and also contains some tightly bound noncarbohydrate material, including proteins and inorganic salts. The exoskeleton protects the soft body and gives support to the body because all the muscles are attached to the inside of the exoskeleton.

Muscular System

The muscular system provides movement internally and externally. Muscles vary in strength and function. Muscles contract and relax to cause movement. Organisms require movement for such functions as obtaining food and oxygen and eliminating wastes. In crawfish, the large abdominal flexor muscle is eaten by humans.

Digestive System

The digestive system converts feed into a form that can be used by the body for maintenance, growth, and reproduction. It consists of all the parts of an organism involved in taking food into the body and preparing it for incorporation into the body. In its simplest form, the digestive system is a tube extending from the mouth to the anus with associated organs. In most species this includes the mouth, esophagus, stomach, intestines, anus, and other associated organs, such as the liver.

Excretory System

Life processes produce waste products. The excretory system eliminates wastes from the body. Typically it consists of the kidneys, urinary ducts, urinary bladder, and urinary opening. Kidneys filter the wastes from the blood. The urinary bladder holds the wastes until they are excreted through the urinary opening.

Respiratory System

The respiratory system takes in oxygen from the environment, delivers it to the tissues and cells of the body, and it picks up carbon dioxide from the tissues and cells, delivering it to the environment. Gills are the respiratory organs of crustaceans. Water is moved over the gills, where oxygen is removed by diffusion into the blood. In the crawfish, gills are located at the base of the legs or maxillipeds and protected by the thorax or carapace. The gills are exposed to water every time the legs or maxillipeds move.

Circulatory System

The circulatory system distributes blood throughout the body. This system consists of a simple heart, large arteries, and sinuses. In the crawfish, the colorless blood is pumped by a very simple heart into several large arteries that pour the blood over the major organs. Then the blood collects in spaces called sinuses and eventually returns to the heart. Blood pumped to the gills picks up oxygen and carries it to the rest of the body. Oxygen is necessary for all cells of the body. As the blood delivers oxygen to the cells of the body, it picks up carbon dioxide, a waste product, which is carried in the blood back through the veins to the heart and gills. The gills release the carbon dioxide to the environment and pick up more oxygen. The blood also delivers nutrients to tissues and cells and removes metabolic wastes from tissues and cells.

Nervous System

The nervous system supplies the body with information about its internal and external environment. This system conveys sensation impulses—electrical-chemical changes—between the brain or spinal cord and other parts of the body. It consists of the brain, the ventral nerve cord, the ganglia, many nerve fibers, and the sensory receptors. The sense organs or receptors receive stimuli and convey these by the nerve fibers to the brain or ventral nerve cord, where they are interpreted. The brain or ventral nerve cord may send responses to the stimuli back through the nerve fibers.

Sensory System

The sensory system includes the five senses—sight, touch, taste, smell, and hearing. The sensory system relays information through the nervous system. The acute senses of smell and touch are located in the antennae, maxillae, and maxillipeds. The compound eyes are on movable stalks, and sight is probably not keen. Hearing is poorly developed, but ear sacs located at the base of the antennules probably aid in balance.

Reproductive System

Sexual reproduction is the process of creating new organisms of the same species through the union of the male and female sex cells—sperm and eggs. Crustacean life cycle and reproduction are quite complex. Testes of males and ovaries of females are located inside the exoskeleton. A duct from the testes or ovaries leads to the outside for the release of sperm or eggs. Using one of the appendages, the male deposits sperm into a receptacle on the female's abdomen.

Members of the pandalids group of shrimp all begin life as males. At about two years of age, they change to females.

MATERIALS

➤ Crawfish—fresh, frozen, or preserved

➤ Forceps

➤ Dissecting tray

➤ Dissecting scissors

➤ Magnifying glass

➤ Blunt probe

PROCEDURES

The crawfish will be observed externally before being dissected.

External Anatomy

Figure 6-1 shows the external anatomy of the crawfish, a representative crustacean. Use Figure 6-1 to identify the following regions and structures:

1. Before identifying any structures locate the following surfaces or regions:
   - Anterior
   - Posterior
   - Dorsal
   - Ventral
   - Lateral

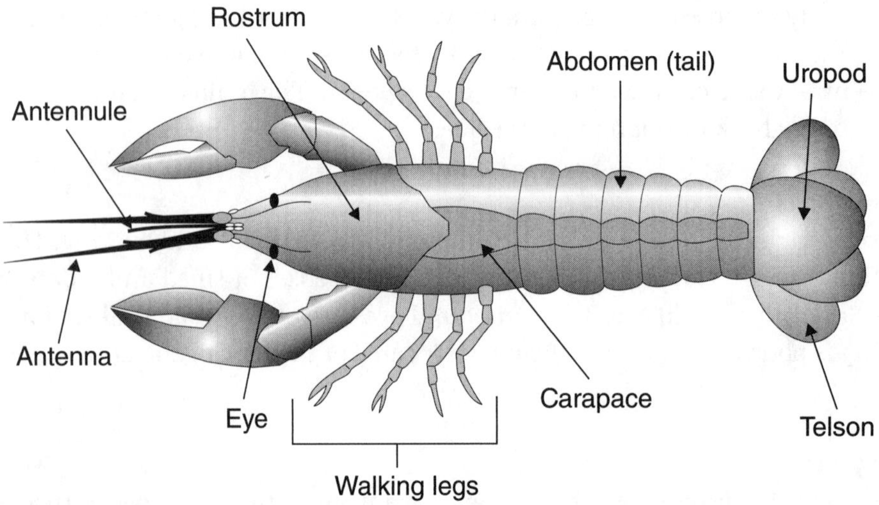

FIGURE 6-1 External anatomy of the crawfish.

2. Identify the three sections of the crustacean body:
   - Head
   - Thorax (carapace)
   - Abdomen

3. Note that each segment of the crustacean body has a pair of appendages.

4. Locate the two pairs of antennae on the head.

5. Next to the antennae, identify the mandibles or true jaws and then two pairs of maxillae or little jaws, which aid in chewing food. The jaws work from side to side, not up and down.

6. Identify the first appendages of the thorax—three pairs of maxillipeds or jaw feet. These hold food during chewing.

7. Next, identify the large claws, or chelipeds, for protection and food getting.

8. Identify the last four pairs of legs on the thorax. Note the two pairs with tiny pincers at the tip and two more pairs with claws.

9. Locate the abdominal appendages of the crawfish. These are called swimmerets or pleopods and are small on the first five segments. In the female during reproduction, the eggs attach to her swimmerets.

10. Locate the sixth swimmeret, which develops into a flipper, or uropod, for locomotion. The telson is attached to the uropod and aids in locomotion.

Internal Anatomy

Figures 6-2 and 6-3 illustrate the internal anatomy of a crawfish.

1. Cut away the carapace, which covers the gills. Be careful not to disturb the gills in the chamber. The entire gill system should be exposed undisturbed. Starting at the anterior-most gill, note how the gills are arranged.

2. With scissors, make a midline incision beginning at the posterior dorsal edge of the carapace and extend it to the rostrum. The cut should be very shallow, in order that the muscles and other internal organs are not disturbed.

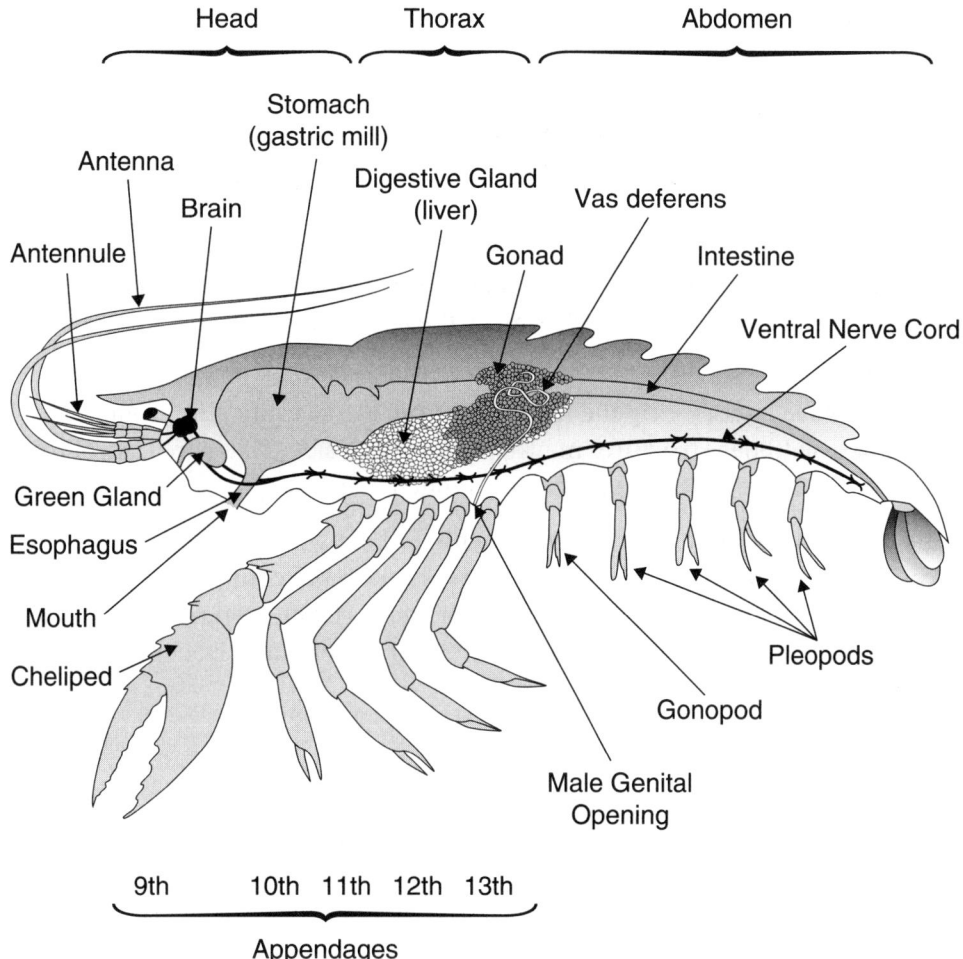

FIGURE 6-2 Internal anatomy of the crawfish, side view.

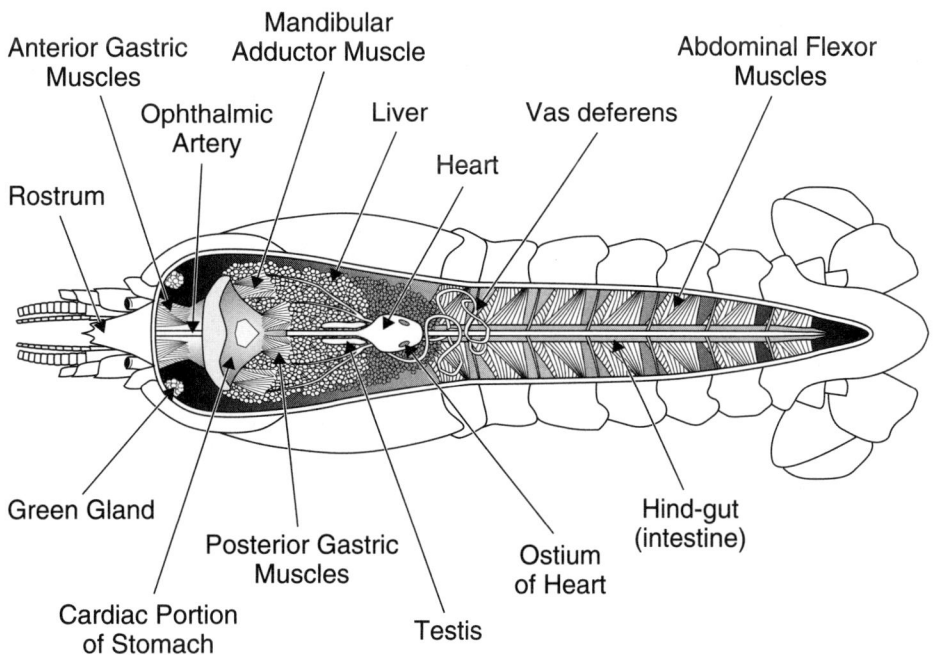

FIGURE 6-3 Internal anatomy of the crawfish, dorsal view.

3. By use of forceps and scissors, remove the carapace.

4. Identify the stomach.

5. Identify the heart and major blood vessels.

6. Remove the right lateral wall of the cephalothorax to the base of the legs to see the liver.

7. Locate the green gland.

8. Locate the gonads. If the specimen is a female, the ovary will be found directly in front of and slightly below the heart. If the sex is male, the testes will be located in about the same position.

9. Remove the stomach by severing the esophagus from the intestine. In the forward, lower portion of the stomach, flattened structures may appear on each side. These are gastroliths. They occur seasonally and will not be present in recently molted individuals.

10. With the stomach and other viscera in the cephalothorax removed, locate the subesophageal ganglion. Extending around the esophagus are single nerve connectives that communicate with the supraesophageal ganglion. Follow the nerve tract posteriorly to the abdomen. This will require careful dissection as the nerve trunk is closely associated with the inner surface of the exoskeleton.

ANALYSIS

1. Compare the size of the carapace to the abdomen. Which part of its body is usually eaten? What does this say about selecting the best species to cultivate?

2. What type of feeding behavior is this crustacean designed for?

3. What opens and closes the chelipeds?

4. Why are crawfish called decapods?

5. Make a list of crustaceans that are closely related to crawfish and used for human food. Identify features that these crustaceans have in common.

6. How are the gastroliths related to the process of molting in crawfish?

7. Observe the maxillipeds through a dissecting microscope or magnifying glass. What observations were made?

8. Describe the eyes of the crawfish.

LAB 7 — *Anatomy of the Mollusca*

INTRODUCTION

Mollusks are separated into classes on the basis of their symmetry, the nature of the foot, the mantle, and the gills. Clams, oysters, scallops, and mussels are bivalves. Their heavy shell does not allow much movement, but it affords maximum protection. The incoming and outgoing siphons that carry water adapt bivalves to a method of trapping food in mucus.

Univalves or gastropods include the snails, conches, and abalones. They have the same general anatomy as the bivalves but only one shell.

The purpose of this lab is to study a representative mollusk—clam, mussel, or oyster.

CORRELATION

This lab and/or Lab 6 can be used with Chapter 7 of *Aquaculture Science*, 3rd Edition.

BACKGROUND

The physiology, or function, of aquatic animals, occurs in body systems. In aquatic animals, these systems are adapted to the water environment. Nine body systems are found in animals, including aquatic animals. These systems are:

1. Skeletal
2. Muscular
3. Digestive
4. Excretory
5. Respiratory
6. Circulatory
7. Nervous
8. Sensory
9. Reproductive

Two shells completely enclose the animal. These shells are made of a calcareous material that is very hard and resembles limestone. Anterior and posterior adductor muscles hold the shells together.

A muscular, hatchet-shaped foot can extend from between the shells. It is used for locomotion and contains much of the digestive, excretory, and reproductive organs. The mantle overlays the internal organs and secretes the hard shell. The digestive system and nervous system are simple. A simple heart pumps the colorless blood to all parts of the body.

The gills not only serve as a respiratory system, but they filter material from the water that is consumed or discharged. Small particles of matter stick to a thin mucous layer on the gills. The surface of the gills have cilia, small hairlike structures, that continually beat, carrying trapped material to the mouth. Water enters the mollusk through a siphon and passes over the gills. Next, water exits the mollusk through another siphon passing the anus, where undigested matter is excreted.

Mollusks reproduce with eggs and sperms. Typically, eggs are released into the water and fertilized by waterborne sperm. Some bivalves are protandrous, meaning they may change their sex one or more times during their lives. Some mollusks, such as scallops, are hermaphroditic, meaning individual organisms have gonads (testes and ovaries) for both sexes.

MATERIALS

➤ Fresh or preserved specimen—a clam, mussel, or oyster

➤ Magnifying glass

➤ Dissecting tray

➤ Dissecting needle

➤ Blunt probe

➤ Dissecting scissors

➤ Scalpel

PROCEDURES

Use Figure 7-1 as a guide during the dissection.

1. Examine your specimen and note that the shell consists of two parts called valves. The shells are fastened together along the dorsal surface by an elastic ligament.

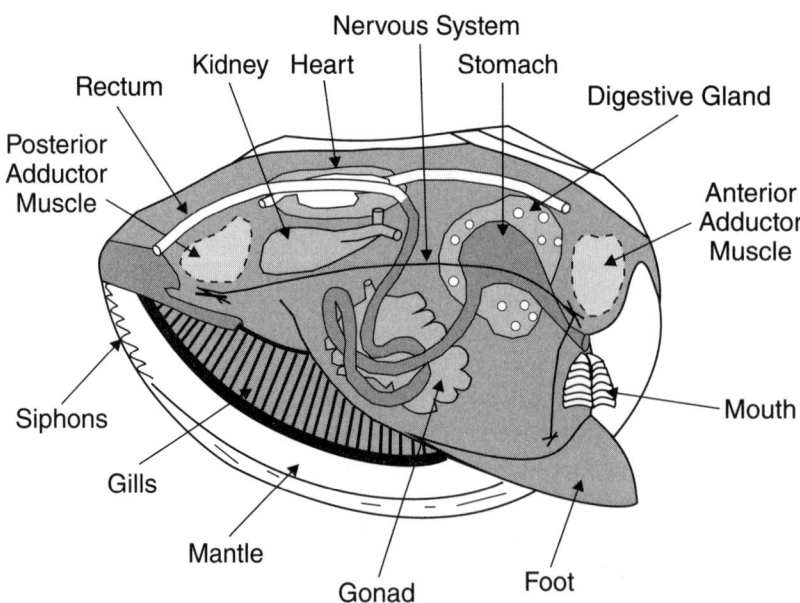

FIGURE 7-1 Internal anatomy of a representative mollusk—a clam.

2. Locate the elevated hump, the umbo, that is near the anterior end of the ligament. This is the oldest part of the shell. As the mollusk grows, it secretes additional layers, each one extending beyond the last one laid down. This produces a series of concentric annual growth rings, which you can locate on the shell.

3. Locate the following surfaces or regions on the mollusk:
   - Dorsal
   - Ventral
   - Posterior
   - Anterior
   - Lateral

4. Look for an outer portion of the shell that is a thin, dark, horny layer. It is often lacking on the umbo of older specimens. The middle layer, or prismatic layer, is formed of crystals. Examine the inner layer, or pearly layer.

5. Using your probe, locate the two openings on the posterior end. The ventral one is an incurrent opening for the entrance of water currents, and the dorsal one is the excurrent opening through which currents of water leave the clam. The shells open on the ventral surface.

6. Examine the outer shell of the bivalve and take note of any byssal threads.

7. Use a flat instrument to open the shells.

8. Cut the two large muscles that hold the shell closed. These are the anterior and posterior adductor muscles.

9. Open the shell and lay it back on the dissecting tray. Note the scars that are left by the removal of the muscles.

10. Examine the dorsal edge of the shell and note any toothlike projections that fit into the grooves in the opposite shell.

11. Locate the visceral mass that lies in the dorsal region between the two large adductor muscles.

12. Examine the mantle covering the visceral mass. Note that it extends ventrally, forming two mantle lobes, one just inside each shell. The space between the two lobes is the mantle cavity.

13. Remove a mantle lobe and expose the mantle cavity and its organs. The muscular foot extends into the mantle cavity and out between the shells.

14. Locate the pair of sheetlike gills that hang into the mantle cavity on either side of the foot.

15. Use the magnifying glass to find the cilia on the gills. The beating of these cilia draws water into the incurrent opening.

16. Locate the mouth and the pair of folds, the palps, on either side of the stomach.

17. Insert the blunt probe into the mouth and direct it into the esophagus and into a saclike stomach.

18. Use a scalpel or scissors to cut away the foot muscle and expose the intestine, digestive gland surrounding the stomach, and the pericardial cavity containing the heart. The digestive gland is usually a green, pulpy tissue. The yellowish to cream-colored tissue surrounding the intestine is mostly the gonad and some muscle fibers.

ANALYSIS

1. What is the shell made of? What does this say about the water chemistry requirements?

2. What are the byssal threads used for?

3. When the stomach and gills were cut open, was any green or brown pigmented debris noted? What are the gills used for? What is this debris?

4. Why do bivalves appear to be a good aquaculture species? Consider food and feeding habits.

5. Compare the mobility of the adult oyster, clam, scallop, mussel, and snail.

6. Draw the general shape of six mollusk shells. Label the shell with the name of the mollusk.

7. If possible, break a small piece of shell and place it into hydrochloric acid and place another small piece into sodium hydroxide. What happens? What does this indicate about the composition of the shell?

INTRODUCTION

Hatching frog eggs and raising larval amphibians through metamorphosis can be a fascinating and educational experience. (See Figure 8-1.)

Students will learn about the metamorphic changes that frogs undergo in their life cycle, and students will understand the habitat and food needs of tadpoles and frogs.

CORRELATION

This lab can be used with Chapter 8 of *Aquaculture Science*, 3rd Edition.

BACKGROUND

Frogs are selective and only breed when temperatures in the air and water are just right. This means that their breeding times are not exact and will be different each year. To make things a little more interesting, the number of eggs laid, as well as their shape and size, also vary between species. The process of metamorphosis varies for each species as well. Table 8-1 is a guide to toad and frog egg identification and timing of events in the lives of frogs to guide this activity.

Amphibians tend to release hormones that inhibit the growth of other amphibians in the tank. This results in pinhead-sized tadpoles sharing the same space with larger emergent frogs. To prevent this, do not overcrowd. Release extras, or get multiple tanks to handle the extras.

FIGURE 8-1 Adult bullfrogs being cultured for frog legs and scientific investigations.

TABLE 8-1 FROG AND TOAD IDENTIFICATION AND BIOLOGY

Name	Identification	Habitat	Breeding	Egg Identification	Metamorphosis
American Toad	2 to 4½ in. (5.1 to 8.9 cm); brown to red to olive; dark, warty skin; elongated glands found at the ridge behind the eye or connected by a short spur.	Common in a variety of habitats wherever there are insects, moisture, and a variety of shallow waters for breeding.	April through June.	Found in long strands 10 to 15 mm wide and several meters long with 4,000 to 8,000 eggs.	Eggs hatch into tiny black tadpoles in 2 to 8 days. Metamorphosis occurs in late June or early July (roughly 2 months), maturity takes 2 to 3 years.
Bullfrog	3½ to 6 in. (9 to 15 cm); plain green with dark markings; no ridges along its back.	Permanent bodies of water.	June through July.	Black and white colored eggs, 1.2 to 1.8 mm diameter, found in masses of 1,000 to 5,000 eggs. Masses 60 cm × 30 to 60 cm; laid as a large surface film several layers thick.	Tadpoles metamorphose in July and August of their second year. Young frogs take 2 to 3 years to mature.
Cope's Gray Treefrog	1¼ to 2 in. (3.2 to 5.1 cm); green to gray to brown; light spot beneath eye; bright yellow or orange on concealed surfaces of hind legs; large toe pads.	Trees or shrubs growing in or near water.	May through mid-July.	Females deposit 10 to 40 eggs at 1.1 to 1.2 mm diameter in loose clusters attached to plants or floating on the water surface.	Not available.
Eastern Gray Tree Frog	1¼ to 2 in. (3.2 to 5.1 cm); green to gray to brown; light spot beneath eye; bright yellow or orange on underside of hind legs; large toe pads; a rough or bumpy skin on its back, usually with darker blotches.	Trees or shrubs growing in or near water.	May through mid-July.	Up to 2,000 eggs singly or in loose clusters of up to 30 eggs attached to vegetation near the surface.	Eggs hatch in 3 to 6 days and tadpoles metamorphose in 6 to 8 weeks.
Green Frog	2¼ to 3½ in. (5.7 to 8.9 cm); green to brown; ridges along its back that do not reach groin; green on upper lip.	All types of permanent bodies of water.	June through July.	Egg mass thin film 30 × 30 cm with 1,000 to 4,000 eggs; 1.0 to 1.5 mm each in diameter. Egg mass floats on the surface.	70 to 85 days.
Mink Frog	17/8 to 2¾ in. (4.8 to 7.0 cm); olive to brown often with spots or mottling on the sides and legs; skin produces a musky, minklike odor when rubbed.	Cool, permanent water where vegetation is abundant (including bogs).	June through July.	2,000 to 4,000 brown-black eggs laid in a loose globular mass; 75 × 125 mm, attached to submerged plants at about 1 meter below the surface.	One year.

TABLE 8-1 FROG AND TOAD IDENTIFICATION AND BIOLOGY (*Continued*)					
Name	**Identification**	**Habitat**	**Breeding**	**Egg Identification**	**Metamorphosis**
Northern Leopard Frog	2 to 3½ in. (5.1 to 8.9 cm); green or brown; rounded spots with light borders; light stripe on upper lip; ridges on its back that extends to groin.	Lakes, streams, rivers, ponds; often far from standing water.	April through mid-June.	Black eggs in a tight globular mass, 10 to 16 cm in diameter, of up to 6,000 in number; 1 to 1.8 mm; eggs laid several centimeters under water on vegetation.	Eggs hatch in 13 to 20 days. Metamorphosis occurs in late June to mid-July, approximately 70 to 100 days.
Northern Spring Peeper	¾ to 1¼ in. (1.9 to 3.2 cm); tan to brown to gray; a dark, often imperfect "X" on the back; a plain belly; large toe pads.	Wooded areas with temporary or semipermanent ponds or swamps or marshes.	March through May.	Egg masses have 800 to 1,000 small white eggs, 1 mm diameter singly or clustered in 2 or 3 attached to grass or vegetation.	Eggs hatch in 2 to 3 days. Tadpole transformers in late May to early June when 8 to 15 mm long.
Pickerel Frog	1¾ to 3 in. (4.4 to 7.6 cm); brown or tan; rectangular spots, without light borders, in parallel rows down the back; bright yellow or orange on concealed surfaces of hind legs; light stripe on upper lip; ridges along its back which extend to the groin.	Cool, clear waters of spring-fed lakes and streams.	April through mid-June.	Light brown, 2 mm diameter, laid near the water surface on vegetation; found in loose globular masses 90 to 100 mm diameter.	Tadpoles hatch in 12 to 18 days and metamorphose in 60 to 80 days.
Wood Frog	13/8 to 2¾ in. (3.5 to 7.0 cm); pink, tan, or dark brown; dark mask through the eye; prominent ridges on its back; light stripe on upper lip.	In or near moist wooded areas.	March through April.	Large globular mass, 60 to 100 mm diameter, attached to submerged plants, near water surface; 500 to 800 eggs per mass; dark colored.	Eggs take 15 to 20 days (depending on water temperature) to hatch. Tadpoles metamorphose in 1½ to 2 months.

MATERIALS

➤ Small dip nets

➤ Clean jars, plastic bags, or plastic containers

➤ Cooler (refrigerator)

➤ Aquarium (1 gallon per two tadpoles) with air stone and air pump

➤ Valid fishing license or small game license (varies in different states)

➤ Pond water from collection site

➤ Food sources

➤ Real or artificial aquarium plants, small rocks, or a small piece of driftwood

PROCEDURES

The procedures to complete this lab activity are divided into five major steps.

Step 1. Collect Eggs and Larvae

In some states, collecting frogs (including their eggs and larvae) may require a valid fishing license or a small-game license. Frog season opens on the Saturday nearest May 1 and runs through December 31 each year.

Never remove eggs or larvae from public areas such as parks, refuges, or conservation areas. Ask permission before removing specimens from private land. Only collect a few larvae and eggs and only take as many as your bowl or aquarium can hold without overcrowding (1 gallon per two tadpoles).

Ponds, small lakes, and creeks are ideal places to find amphibian eggs and catch tadpoles. Use small dip nets or jars to collect eggs and larvae and transport them to the classroom in clean jars, plastic bags, or plastic containers. Take a temperature reading of the water. Put the eggs or larvae in an insulated bag or cooler in order to maintain the approximate temperature of the water. Take an extra container of water from the water body where the specimens were collected to place in the aquarium.

Step 2. Set up the Habitat

Eggs and tadpoles can be kept in a large, flat pan, a fish bowl, an aquarium, or a large glass jar, but the habitat needs to be set up ahead of time. Every two tadpoles require at least 1 gallon of water to prevent overcrowding.

Use water from the pond where the eggs or larvae were collected to give them a head start.

Note: Chlorinated tap water destroys bacteria and algae, and it can harm or kill amphibian eggs and larvae. If tap water is used, it often needs to be treated with a dechlorinator. Or, a container of water left to stand a few days with the lid off allows the chlorine to dissipate.

Use of an air stone and air pump provide a constant stream of fine bubbles. Providing sand or gravel is not necessary. Eggs found in submerged habitats should be kept submerged, and those found floating should be allowed to float.

Step 3. Feed the Tadpoles, Frogs, or Toads

Tadpoles usually eat algae and other minute plant matter, but this may be hard to get in sufficient quantities. Finely ground commercial goldfish food, a commercial Trout Chow, or algae from another aquarium should be fed twice daily. As a substitute, boil and cool two tablespoons of fresh spinach or lettuce (not cabbage). Crushed rabbit food pellets can be fed to tadpoles as a dietary supplement. Also, small flakes of hard-boiled egg yolk can be added twice a week as a protein supplement.

Feed only what the tadpoles can eat in an hour to avoid fouling the water. Remove any uneaten food promptly.

As tadpoles become frogs, their diet changes from eating plants to feeding exclusively on live animals such as insects and small crustaceans. Finding enough food to maintain most juvenile frogs for very long becomes a real challenge. Tiny mealworms or aphids from infested houseplants can be used. (**Note:** An additional assignment for other class members could be a project that produces mealworms, aphids, or crickets to feed the frogs.)

Frogs instinctively eat living creatures. These creatures attempt to avoid being caught and try to escape while being eaten. Unless the food in the frog's mouth moves or wiggles, frogs will spit it out. To overcome this instinctive behavior, frogs must be taught to eat an inert (non-living) food during and shortly after metamorphosis. The foods that frogs have been taught to eat include pelletized food, dropping food, and crawfish heads.

When frogs are going through metamorphosis, small pieces of crawfish bodies are wiggled in front of them. They grab the pieces with their tongues, sometimes eating them and sometimes spitting them out. Worms can also be fed at this stage. This allows the frogs to associate the inert crawfish bodies with food, and they become accustomed to the crawfish.

Frogs can also be trained to eat pelletized food at this stage of development. Small pellets can be dropped in the vicinity of the frogs, and they will grab the food with their tongues.

Step 4. Record Transformation to Adult Stages in a Daily Log

Use Table 8-2 to record daily observations and document how long it takes for each stage to occur as tadpoles become frogs. Tadpoles undergo three remarkable changes that are easy to observe (Figure 8-2)—

➤ Grow legs—back legs first; front legs last

➤ Slowly lose their tails

➤ Switch from breathing with gills to breathing with lungs after growing legs

As the front legs grow, the tadpoles will no longer eat. The tail shrinks as the tissue is reabsorbed as food by the tadpole.

Once the tadpoles' hind legs appear, rework the landscape in their container. Create a habitat design for the tank. Provide a gently sloping place where the froglets can crawl out of the water. When the froglets are ready to leave the water, they must be able to do so quickly, or they may drown. A small pile of rocks is fine. Driftwood also works, but avoid all types of treated lumber.

Step 5. Return Amphibians to their Natural Habitat

After completing observations, release the frog back into the wild where they were originally collected as eggs or larvae.

ANALYSIS

Write a lab report based on the notes made in Table 8-2 log. Respond to the remaining Analysis items in your lab report. Your instructor can provide guidelines for your lab report or the Introduction to this Lab Manual provides some guidelines.

1. Discuss the diet of tadpoles and frogs throughout their life cycle.

2. How can frogs and tadpoles be provided with the specific foods they need?

3. What feeding plan did you devise and schedule for tadpoles and frogs?

4. Discuss how some amphibians migrate to find the habitat they need during the winter, and what type of habitat each one needs.

5. Discuss how amphibians survive during the winter.

6. Identify potential uses and markets for cultured frogs.

TABLE 8-2 DAILY LOG FOR OBSERVING TADPOLES AND FROGS		
Date	Feed	Observations

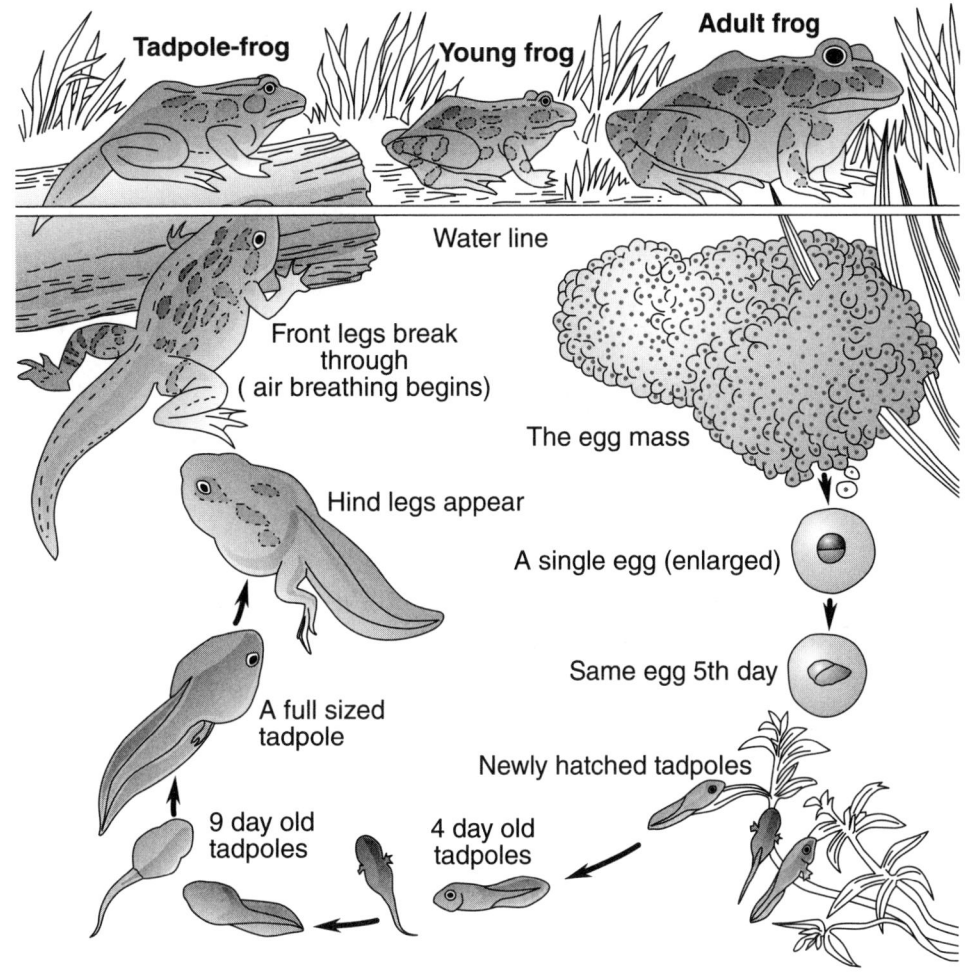

Tadpole-frog

Young frog

Adult frog

Water line

Front legs break
through
(air breathing begins)

The egg mass

Hind legs appear

A single egg (enlarged)

Same egg 5th day

A full sized
tadpole

Newly hatched tadpoles

9 day old
tadpoles

4 day old
tadpoles

FIGURE 8-2 Life cycle of the frog.

SUPPLIERS

Frog kits are available if eggs cannot be collected. These kits come complete with mini-aquariums (habitats) and all the supplies to raise frogs from the egg stage or the tadpole stage. Kits are available from the following companies:

Three Rivers of Brooksville, Inc.
P.O. Box 10369
Brooksville, FL 34603
(352) 544-0333
Fax (353) 848-0100
www.growafrog.com

Carolina Biological Supply Company, Inc.
P.O. Box 6010
Burlington, NC 27215
(800) 334-5551
(http://www.carolina.com)

LAB 9

Nutrients—Proteins, Carbohydrates, and Fats

INTRODUCTION

Like traditional livestock, fish derive their energy from three sources—carbohydrates, fats, and proteins. Because carbohydrates are used rather inefficiently in most fish systems, the primary energy sources are fats and proteins. Proteins make up the bulk of all solid material within the bodies of animals. They also make up some hormones that function in chemical control in the body.

The purpose of this lab is to demonstrate some of the simple tests for the presence of proteins and fats and some of their characteristics.

CORRELATION

This lab can be used with Chapters 9 and 10 of *Aquaculture Science*, 3rd Edition.

BACKGROUND

The primary purpose of aquaculture is the efficient conversion of feed into meat for consumption. Fish convert feed to food for human consumption very efficiently, especially compared to other meat-producing animals. The digestive system acts like an assembly line in reverse, taking the feedstuffs apart into their basic chemical components so that the fish can absorb them and rearrange them into its own characteristic body composition. The major feedstuff groups are carbohydrates, proteins, and fats.

Carbohydrates

Fish digest simple sugars efficiently. (See Figure 9-1.) As the sugar becomes larger and more complex—like starch—digestibility decreases rapidly. Warmwater fish digest dietary carbohydrates better than coldwater

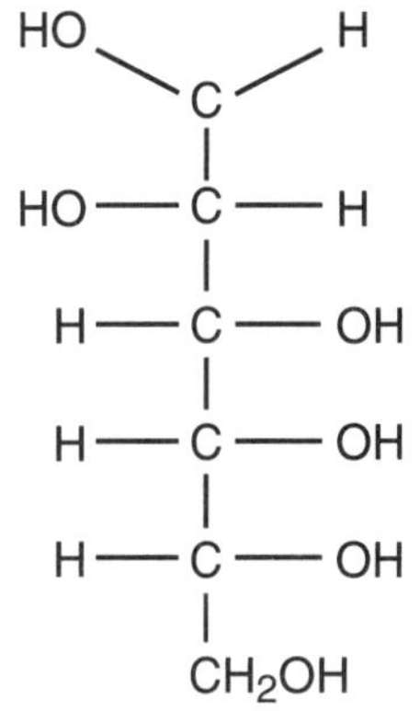

FIGURE 9-1 Chemical structure of the simple sugar glucose.

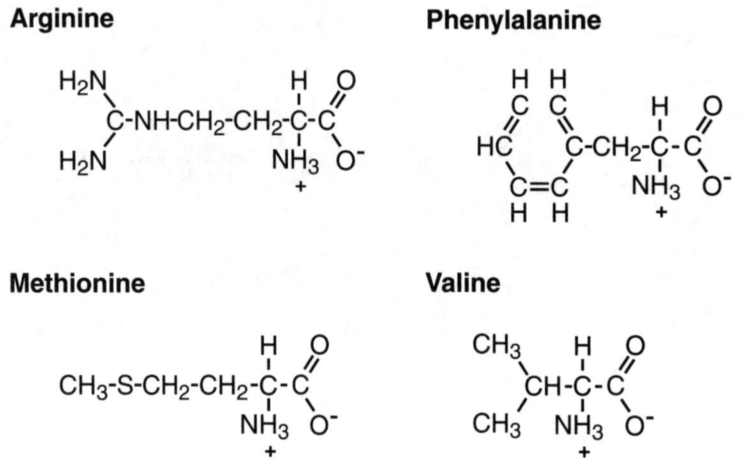

Arginine

Phenylalanine

Methionine

Valine

FIGURE 9-2 Chemical structure of some essential amino acids.

or marine fish do. The ability to use carbohydrates as an energy source varies among the fish species. Carbohydrates improve growth, provide precursors for some amino acids, and nucleic acids. Also, carbohydrate is the least expensive source of dietary energy. In warmwater fish, cereal grains provide inexpensive sources of carbohydrates, but their use is limited in coldwater fish. Digestible carbohydrates in trout feed are generally lower than the levels in catfish feed. In nutrition, carbohydrates spare protein because less protein will be used for energy.

Proteins

Proteins are long chains of amino acids linked by bonds called *peptide bonds*. Figure 9-2 shows diagrams of several amino acids. All amino acids contain nitrogen, so all proteins contain nitrogen. In fact, measuring the nitrogen content is a method of calculating protein content. Metabolism of protein for energy produces nitrogen end products. Fish eliminate these through the gills, feces, and urine. These nitrogen end products can cause problems in fish ponds.

Protein serves three purposes in the nutrition of fish:

➤ Provide energy

➤ Supply amino acids

➤ Meet requirements for functional proteins—enzymes, hormones, and structural proteins.

The requirement for protein in fish diets is essentially a requirement for the amino acids in the dietary proteins. Some amino acids the fish cannot synthesize are called *indispensable* or *essential* amino acids. These include 10 amino acids: arginine, histidine, isoleucine, leucine, lysine, methionine, phenylalanine, threonine, tryptophan, and valine.

Fat

Each gram of fat contains two and one-half times the energy in carbohydrates or proteins.

Besides being an important source of energy for fish, dietary fats provide essential fatty acids needed for normal growth and development. Also, dietary fats assist in the absorption of fat-soluble vitamins.

Fish cannot synthesize some fatty acids. These become essential fatty acids (EFA). Freshwater fish require a dietary source of linoleic acid and/or linolenic acid (Figure 9-3). These are both 18-carbon fatty acids.

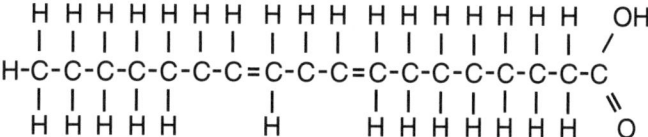

FIGURE 9-3 Chemical structure of the essential fatty acid linoleic acid.

Fish diets are formulated to meet the optimum ratio of energy to protein for each species. Fats serve as an important source of energy, but a definite percentage of dietary fat cannot be given without considering the type of fat, as well as the protein and energy content of the diet.

MATERIALS

- ➤ Glass marking pen
- ➤ Test tubes
- ➤ Test tube rack
- ➤ Test tube stoppers or covers
- ➤ Droppers
- ➤ Nitric acid
- ➤ Iodine solution
- ➤ Fresh potato
- ➤ Tuna fish
- ➤ Oat cereal
- ➤ Egg white (hard-boiled)
- ➤ Cotton balls
- ➤ Shortening
- ➤ Olive, corn, canola, or peanut oil
- ➤ Water
- ➤ Brown paper (paper bag)
- ➤ Alcohol
- ➤ Liquid detergent

PROCEDURES

These are simple tests to identify the presence of starch—a carbohydrate—protein, and fat. These tests are qualitative, but more sophisticated tests that are quantitative also rely on color changes.

Identification of Starch

1. Number five clean test tubes 1 to 5. Place them in a test tube rack. Using Figure 9-4 as a guide, add pea-sized pieces of the following substances to each test tube:

   Tube 1—uncooked oatmeal cereal (rolled oats)

   Tube 2—egg white, hard-boiled

   Tube 3—cotton

   Tube 4—potato

   Tube 5—shortening

2. Add 3 to 5 drops of iodine solution to each test tube.

3. Wait several minutes. Then, record the color of the items placed in each tube in Table 9-1. The test used to identify starch turns starch-containing substances to a blue-black color. No color change to blue-black indicates that the substance being tested contains no starch.

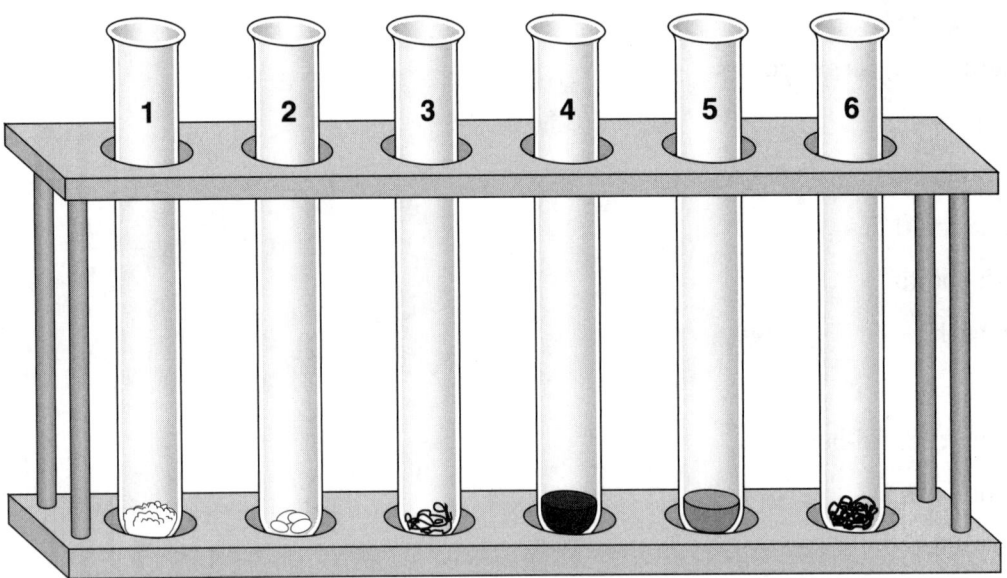

FIGURE 9-4 Numbered test tubes in a rack ready for the addition of iodine.

TABLE 9-1 IDENTIFICATION OF STARCH-CONTAINING SUBSTANCES		
Tube	**Substance**	**Color Change Due to Iodine**
1	Oat cereal	
2	Egg white	
3	Cotton	
4	Potato	
5	Shortening	

TABLE 9-2 IDENTIFICATION OF PROTEIN-CONTAINING SUBSTANCES

Tube	Substance	Color Change Due to Iodine
1	Tuna fish	
2	Egg white	
3	Cotton	
4	Potato	
5	Shortening	

Identification of Proteins

1. Number five clean test tubes 1 to 5. Place them in a test tube rack. Using Figure 9-4 as a guide, add pea-sized pieces of the following substances to each test tube:

   Tube 1—tuna fish

   Tube 2—egg white, hard-boiled

   Tube 3—cotton

   Tube 4—potato

   Tube 5—shortening

2. Add 5 drops of nitric acid to each test tube.

CAUTION: Nitric acid is harmful to skin and clothing. Rinse with water if spillage occurs. Call your teacher.

3. Wait several minutes. Then, record the color of the items placed in each tube in Table 9-2. The test used to identify protein is called the *xanthoproteic* test. A substance containing protein will turn yellow when nitric acid is added to it. No color change to yellow indicates that the substance being tested contains no protein.

Identification of Fats

Fats can be identified by their solubility and the brown paper test.

1. Number eight clean test tubes 1 to 6. Place them in a test tube rack.

2. Add about 5 ml of cold water to tubes 1 through 4.

3. Add about 5 ml of alcohol to tubes 5 through 8.

4. Add the following substances to each test tube:

   Tube 1—five drops of oil

   Tube 2—five drops of oil and 10 drops of liquid detergent

   Tube 3—grain-sized sample of shortening

   Tube 4—grain-sized sample of shortening plus 10 drops of liquid detergent

   Tube 5—grain-sized sample of shortening

   Tube 6—five drops of oil

   Tube 7—five drops of water

   Tube 8—pea-sized piece of potato

Tube	Observations after Shaking	Results of Brown Paper Spot
TABLE 9-3 RESULTS OF TESTS ON FATS IN SOLUTIONS		
1		—
2		—
3		—
4		—
5		
6		
7		
8		

CAUTION: Alcohol is flammable. Extinguish all flames in the laboratory before proceeding.

5. Mix contents of each tube by placing a stopper over the opening of each tube. Place your thumb over the stopper and shake each tube ten times.

6. Wait 1 minute.

7. Examine and compare all the tubes. Fats are soluble in alcohol. Soluble means that they dissolve or mix. Fats are not soluble in cold water. They do not dissolve or mix. If the fat is an oil, two layers will be seen. Detergents act like the bile in the digestive tract and allow the fats to break down and mix with water.

8. Record your observations in Table 9-3.

9. On separate pieces of brown paper numbered 5 through 8, pour a small amount of the solution from the corresponding test tube.

10. Allow the paper to dry for a few minutes.

11. Hold the paper toward light. If light passes through, a translucent (semitransparent) spot has formed. Examine the pieces of paper to check for a translucent spot. Record the results in Table 9-3. Fats should produce a translucent spot.

ANALYSIS

1. If iodine does not turn a food bluish-black, does this tell you what type of food it is?

2. What is the difference between starch and sugars?

3. Compare Figures 9-1, 9-2, and 9-3. Discuss the differences and similarities of carbohydrates, proteins, and fats.

4. In Latin, the word *xantho* means yellow and *proteic* means protein. Why is *xanthoproteic* a meaningful word to use when describing the chemical test used for identifying a protein?

5. If nitric acid does not turn a food yellow, does this tell you what type of food it is?

6. Why is it important in nutrition to know which foods contain protein, which contain carbohydrates, and which contain fats?

7. Describe the solubility of fat in water, alcohol, and water with detergent.

8. Does a potato contain fat? Describe how you reached this conclusion.

9. The nitric acid test indicates the presence of protein. Nutritionists need to know how much protein is in a certain feed. They also need to know what amino acids are present in a feed. Why?

10. Based on your observation of fat in a solution with detergent, why is bile necessary for fat digestion?

LAB 10 *Bacteria*

INTRODUCTION

Bacteria are microscopic one-celled organisms having various shapes. They often cause disease, though some bacteria are beneficial. Bacterial diseases are often internal infections and require treatment with medicated feeds containing antibiotics approved for use in fish by the Food and Drug Administration. Typically, fish infected with a bacterial disease will have hemorrhagic spots or ulcers along the body wall and around the eyes and mouth. They may also have an enlarged, fluid-filled abdomen and protruding eyes. Bacterial diseases can also be external, resulting in ulceration and skin erosion.

The purpose of this lab is to acquaint the student with the form and structure of some common bacteria and their names.

CORRELATION

This lab can be used with Chapter 11 of *Aquaculture Science*, 3rd Edition.

BACKGROUND

A bacterium is a microscopic single-celled organism, and very different from a virus. The plural form of *bacterium* is *bacteria*. Bacteria occur everywhere life exists. They possess a tough, rigid outer cell wall through which they absorb their food. Some bacteria have slimy outer capsules, and some may have a whiplike flagella to propel them through liquids. If flagella are positioned all around the bacterium, it is called *peritrichous*, but if flagella are at each end it is called *lophotrichous*. Some bacteria may simply drift in air or water currents.

Bacteria generally reproduce by splitting into two. This is called binary fission, and it may occur once every 15 to 30 minutes. Under favorable conditions, one bacterium could form over 150 trillion bacteria in 24 hours! This usually does not happen. Bacteria are very numerous and very tough. A pinch of soil contains millions, and some bacteria can survive freezing, intense heat, drying, and some disinfectants. To survive adverse conditions, bacteria form spores that can remain active for years.

Bacteria can be classified by their shapes. Bacteria shaped like a sphere are called *cocci*. Those shaped like rods are called *bacilli*. *Spirillum* bacteria are a spiral shape. *Vibrio* are comma-shaped. A *Mycobacterium* is made up of very small rods. *Flexibacter* form long, thin rods. Figure 10-1 illustrates some of the common shapes.

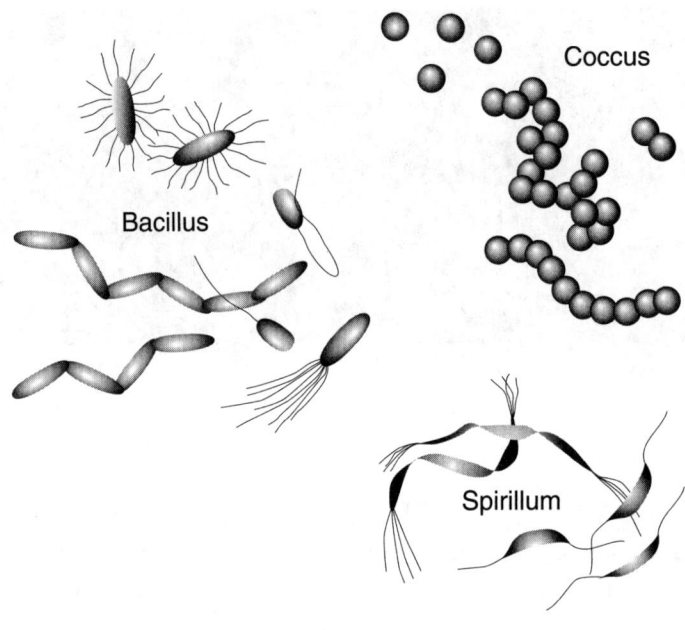

Bacteria

FIGURE 10-1 Examples of some common bacterial shapes.

Many bacteria perform useful functions for humankind. For example, helpful bacteria include those responsible for decay, sewage treatment, cheese and yogurt production, and those responsible for the nitrification process. Some bacteria cause disease. These are called *pathogenic*. Bacterial infections can be treated with antibiotics or similar drugs, and some can be prevented by vaccination.

MATERIALS

➤ Pencil and paper

➤ Prepared microscope slides of bacteria

➤ Microscope

PROCEDURES

1. Using the key in Table 10-1, identify the bacteria in Figure 10-2.

2. Using the prepared microscope slides of bacteria, view ten slides of rod, round, and spiral bacteria.

3. Draw and describe what you see through the microscope. Use Table 10-2.

TABLE 10-1	KEY TO IDENTIFY BACTERIA	
1.	If the general shape of the bacterium is round	Go to 4
2.	If the general shape of the bacterium is rod	Go to 5
3.	If the general shape of the bacterium is spiral	Go to 6
4.	If in pairs	Go to 4a or 4b
	If in chains	Go to 4c or 4d
	If in clumps	*Staphylococcus aureus*
4a.	Without a heavy cover	*Diplococcus meningitidis*
4b.	With a heavy cover (capsule)	*Diplococcus pneumoniae*
4c.	Large in size	*Streptococcus pyogenes*
4d.	Small in size	*Streptococcus lactis*
5.	If in chains	*Bacillus anthracis*
	If in pairs	*Bacillus lactis*
	If single	Go to 5a
5a.	With hairs—flagella	*Bacillus typhosa*
5b.	With a bulge (spore) in the middle	*Bacillus botulinum*
5c.	With a bulge (spore) at the end	*Bacillus tetani*
6.	Spiral bacterium	*Treponema pallidum*

ANALYSIS

1. Not all bacteria are harmful. Name two examples and tell what they do.

2. Sometimes the disease or effect of the bacterium can be determined by its name. Try matching these names with the disease they cause or their effect.

_____	*Staphylococcus aureus*	A.	Typhoid fever
_____	*Diplococcus meningitides*	B.	Botulism poisoning
_____	*Diplococcus pneumoniae*	C.	Syphilis
_____	*Streptococcus pyogenes*	D.	Anthrax
_____	*Streptococcus lactis*	E.	Tetanus
_____	*Bacillus anthracis*	F.	Spinal meningitis
_____	*Bacillus lactis*	G.	Pneumonia
_____	*Bacillus typhosa*	H.	Tonsillitis
_____	*Bacillus botulinum*	I.	Sauerkraut
_____	*Bacillus tetani*	J.	Buttermilk
_____	*Treponema pallidum*	K.	Boils

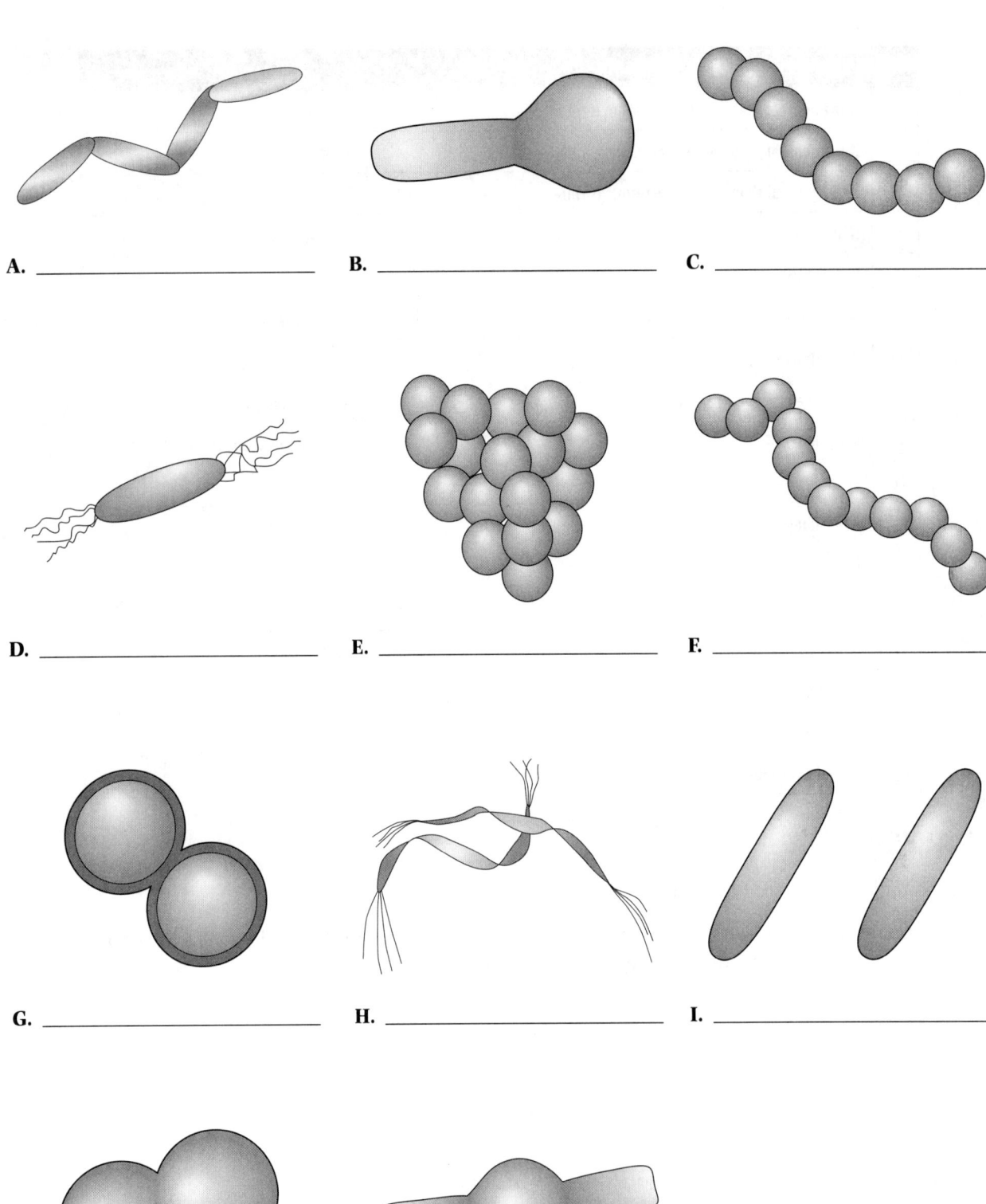

FIGURE 10-2 Examples of some common bacterial shapes.

A. _____

B. _____

C. _____

D. _____

E. _____

F. _____

G. _____

H. _____

I. _____

J. _____

K. _____

TABLE 10-2 DRAWINGS OF BACTERIA VIEWED THROUGH THE MICROSCOPE

Slide _____

Slide _____

Slide _____

Slide _____

Slide _____

Slide _____

Slide _____

Slide _____

Slide _____

Slide _____

3. How do the scientific names of bacteria help describe their features?

4. List some ways to eliminate or at least reduce the numbers of bacteria.

5. What is used to treat fish that have a bacterial disease?

6. How do bacteria and viruses differ?

LAB 11   *Effect of Temperature on Fish Respiration Rate*

INTRODUCTION

Water temperature helps determine which species may or may not be present in the system. Temperature affects feeding, reproduction, immunity, and the metabolism of aquatic animals. Drastic temperature changes can be fatal to aquatic animals. Not only do different species have different requirements, but optimum temperatures can change or have a narrower range for each stage of life.

The purpose of this lab is to determine the influence of water temperature on the respiration rate of goldfish.

CORRELATION

This lab can be used with Chapter 11 of *Aquaculture Science*, 3rd Edition.

BACKGROUND

Cold-blooded animals are literally at the mercy of their environment. Lowering of the body temperature in response to reduced temperature of the environment may result in the crystallization of proteins in the blood, resulting in death. An increase in body temperature may speed up metabolism to such an extent that the tissues are literally burned up.

Temperature stress, particularly cold temperatures, can completely halt the activity of antibodies of the immune system, eliminating an important first defense against invading organisms. A sharp decrease in temperature severely impairs the fish's ability to quickly release antibodies against an invading organism. The time lapse required to mount an antibody response gives the invader time to reproduce and build up its numbers, giving it an advantage that may allow it to overwhelm the fish.

During the respiratory cycle in the fish, water containing oxygen enters the mouth and is forced out over the gill filaments when the mouth is closed. The oxygen dissolved in the water enters the capillaries, and the excess carbon dioxide in the capillaries is released into the water. The operculum opens to allow the carbon-dioxide containing water to leave the gill chamber, completing the respiratory cycle. The counting of the movements of the operculum is one method of computing the respiratory rate of the goldfish.

MATERIALS

➤ Small aquarium, goldfish bowl, or battery jars

➤ Goldfish (comets are best for this experiment)

➤ Thermometer that can be immersed in water

➤ Small dip net

➤ Crushed ice

➤ Stopwatch or watch with second hand

➤ Warm water, about 95°F

➤ Graph paper and pencil

➤ Calculator (optional)

PROCEDURES

1. Put a goldfish in an aquarium, jar, or bowl with sufficient water to cover the dorsal fin.

2. Put a thermometer at one edge of the aquarium so you can read it without disturbing the fish (see Figure 11-1).

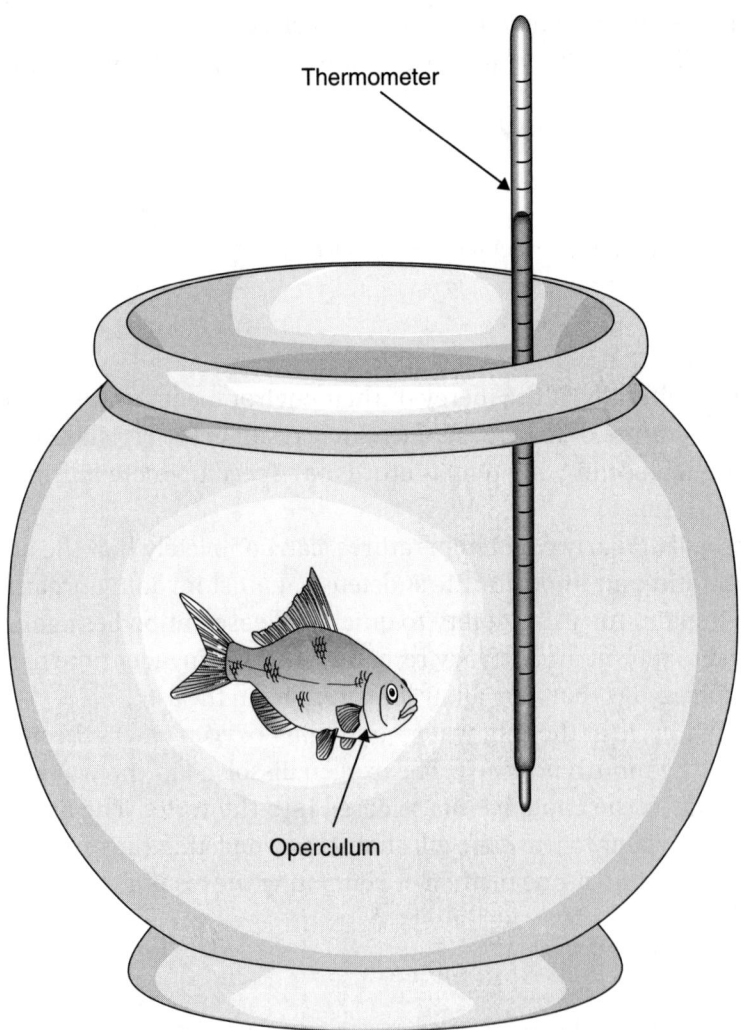

FIGURE 11-1 Counting the movements of the operculum at different water temperatures can be used to calculate the respiration rate in goldfish.

3. In order to reduce the shock factor and avoid exciting the fish, add crushed ice slowly to the aquarium. Continue to add the ice until the water temperature is reduced to near freezing, 32°F.

4. Observe the movement of the operculum covering the gills. Use your watch and count the movements of the operculum at the lowest temperature for one minute. Record this rate in Table 11-1.

5. Slowly add warm water to the aquarium at the end opposite the thermometer. Remove some of the water and ice so that the water volume remains constant. *Avoid exciting the fish when adding the water.*

Continue to add warm water until a 5°F rise in temperature occurs. Wait a few seconds for the fish to adjust to the change, then count and record the rate of operculum movement at this temperature.

6. Continue adding warm water to the aquarium, remembering that some water will have to be removed, and record the rate at each five-degree temperature interval until the water temperature reaches 92°F.

7. After the experiment, remove the goldfish and place it in an aquarium at room temperature and well aerated.

8. Repeat the process with two or three more goldfish and record the results in Table 11-1.

9. Construct a graph showing the respiration rate for each fish at each temperature reading. Calculate the average rate of operculum movement for the goldfish, record in the table provided, and construct a graph based on the averages.

TABLE 11-1 RESPIRATION RATE IN GOLDFISH AT VARIOUS TEMPERATURES

Degrees Fahrenheit	Fish #1	Fish #2	Fish #3	Fish #4	Average Rate
32					
37					
42					
47					
52					
57					
62					
67					
72					
77					
82					
87					
92					

ANALYSIS

1. At which temperature did the greatest number of operculum movements occur?

2. What is the relationship between the increase in water temperature and respiration rate in the goldfish?

3. List the factors that, in addition to water temperature, could influence the respiration rates in your goldfish.

4. Explore other factors such as the effect of heat or extremely cold temperatures on the enzymes involved in respiration, or the effect of oxygen starvation on internal tissues of the fish.

5. Fish are adapted for an aquatic life and live in both warm and cold waters. In terms of your results, discuss how the respiration rates of fish in these environments could vary.

6. Explain why it is necessary to avoid exciting the fish when adding water.

LAB 12

Testing Water: pH, Dissolved Oxygen, Carbon Dioxide, Ammonia, Nitrite, and Hardness

INTRODUCTION

High-quality water and plenty of it is the primary consideration for any aquaculture facility. This is true for finfish, shellfish, and crustaceans. Water provides oxygen and food, serves as an excretory site, helps regulate body temperature, and may harbor disease-causing organisms. Gaining an understanding of water helps aquaculturalists become more productive.

This lab is designed to:

➤ Determine pH of water samples

➤ Measure and record the amount of dissolved oxygen and carbon dioxide in water samples

➤ Determine ammonia and nitrite concentrations of water samples

➤ Measure the hardness of water samples

CORRELATION

This lab and/or Lab 13 can be used with Chapter 12 of *Aquaculture Science*, 3rd Edition.

BACKGROUND

Water quality test kits for aquaculture are frequently used to determine the dissolved oxygen, carbon dioxide, pH, ammonia, nitrite, and hardness of water. This information helps the aquaculturalist make correct management decisions.

pH

The pH—one of the most common water tests—is a measure of hydrogen ions in the water. The pH scale spans a number range of 0 to 14 with the number 7 being neutral. The pH scale is logarithmic, so every one-unit change in pH represents a tenfold change in acidity. Measurements above 7 are basic and below 7 are acidic. The farther a measurement is from 7, the more basic or acidic is the water. Acid and alkaline (basic) death points for fish are approximately pH 4 and pH 11 growth and reproduction can be affected between pH 4 and 6 and pH 9 and 10 for some fishes. Also, pH affects the toxicity of other substances such as ammonia and nitrite.

Dissolved Gases

Dissolved gases determine the basic suitability of water for fish survival. These gases include oxygen (O_2), carbon dioxide (CO_2), nitrogen (N_2), ammonia (NH_4^+ and NH_3), hydrogen sulfide (H_2S), chlorine (Cl_2), and methane (CH_4). Dissolved gases are usually not found in water analysis report forms because the manner by which samples are collected and shipped can cause gas measurements to be much unlike the actual on-site water condition. Ammonia is the most stable of the group, and if a sample is processed within a day after collection, it should measure fairly accurately. Other measures are best taken at the water site using appropriate meters or chemical test procedures.

Dissolved Oxygen. Aquatic life requires dissolved oxygen. It varies greatly in natural surface water and is characteristically absent in groundwater. Most aquatic animals need more than a 1 ppm concentration for survival. Depending on culture circumstances, aquatic animals need 4 to 5 ppm to avoid stress. Concentrations considered typical for surface water are influenced by temperature but usually exceed 7 to 8 mg/l (ppm). In ponds, dissolved oxygen fluctuates greatly due to photosynthetic oxygen production by algae during the day and the continuous consumption of oxygen due to respiration. Dissolved oxygen typically reaches a maximum during the late afternoon and a minimum around sunrise. Cloudy weather, rain, plankton die-offs, and heavy stocking and feeding rates result in low levels of dissolved oxygen, which can stress or kill fish.

Oxygen is only slightly soluble in water. Water may be frequently supersaturated with oxygen in ponds with algae blooms.

As water warms (see Figure 12-1), is raised to higher altitudes (see Figure 12-2), or becomes more saline, its oxygen holding capacity declines. Water saturated with oxygen at 59°F (15°C) contains about 9.8 ppm, while water at 86°F (30°C) is saturated at about 7.5 ppm.

Aquaculturalists measure dissolved oxygen with oxygen meters or chemical test kits that give results in mg/l. Guidelines for oxygen management usually report that oxygen levels should be maintained above 4 mg/l (ppm) to avoid stress. Most warmwater fish experience significant oxygen stress at levels of 2 mg/l, and levels of less than 1 mg/l (ppm) may result in fish kills.

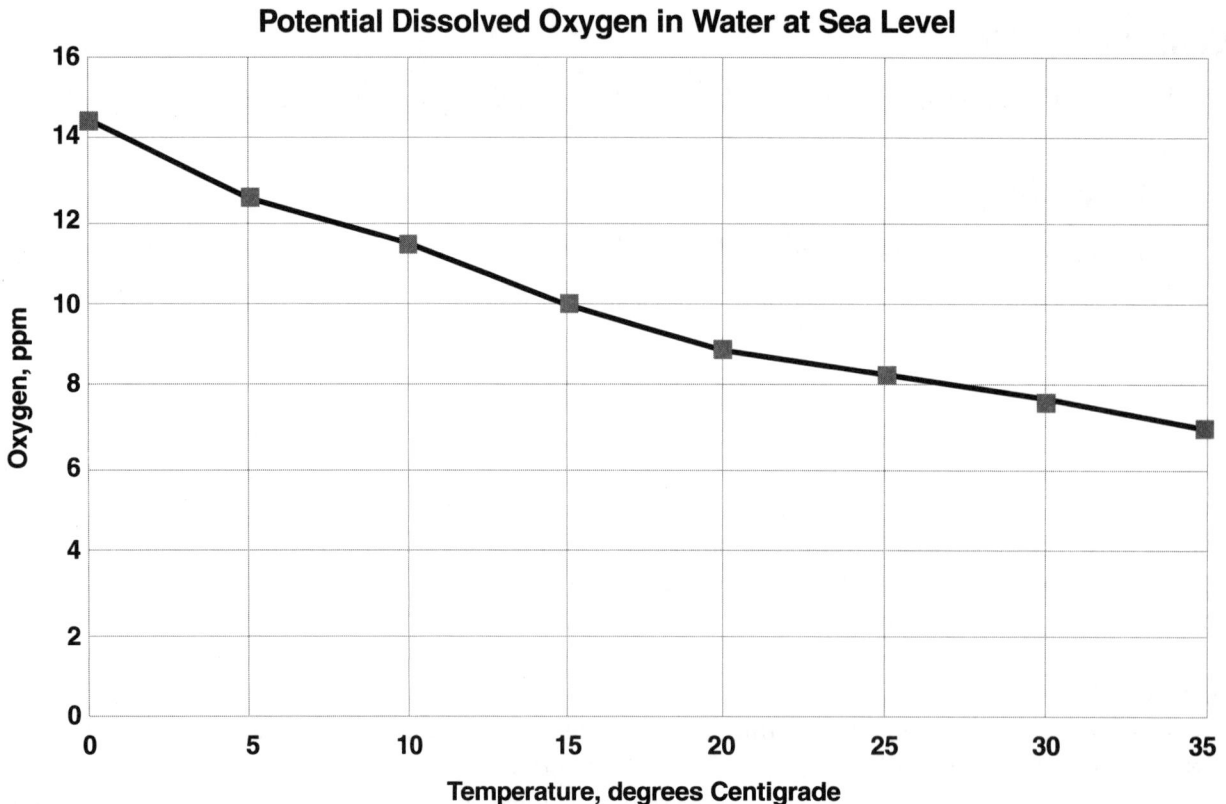

FIGURE 12-1 As water temperature increases, oxygen content decreases.

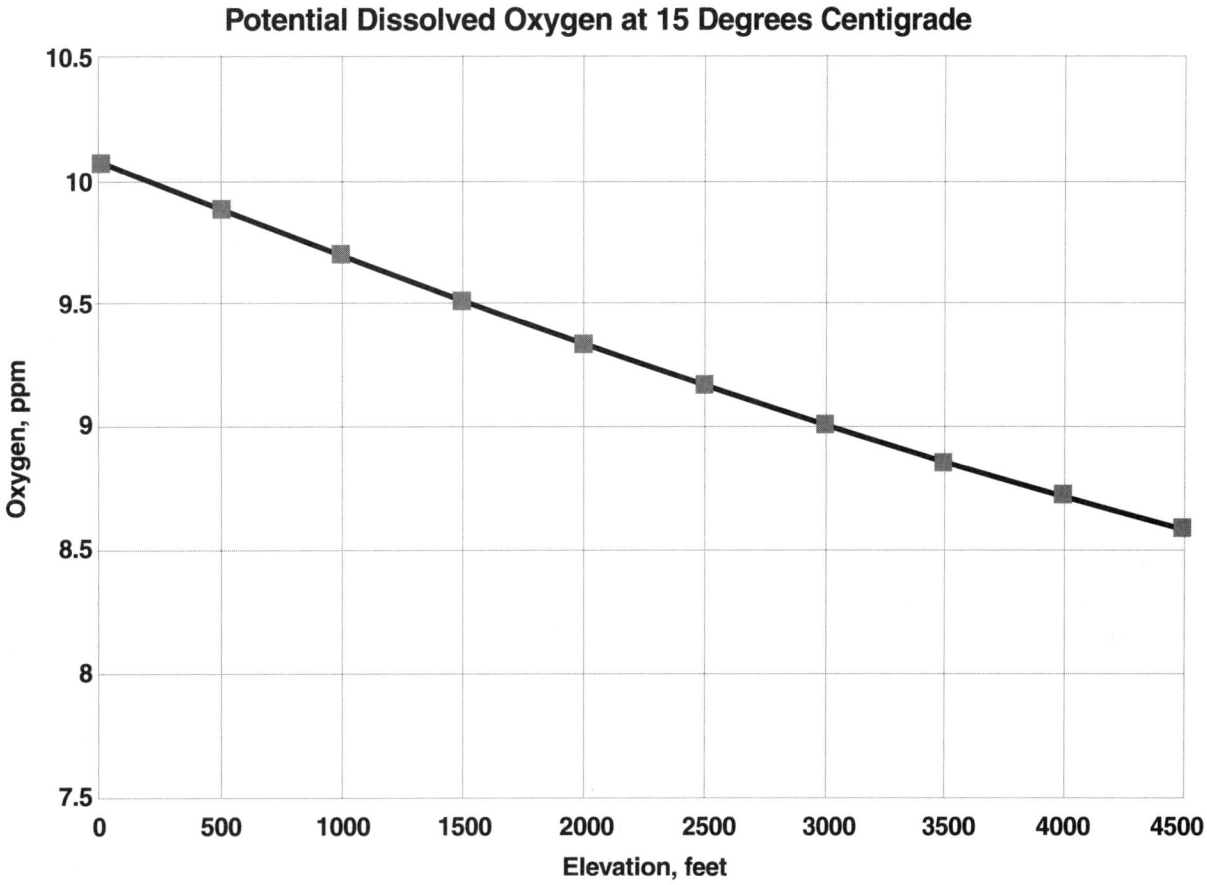

FIGURE 12-2 Water at higher altitudes holds less oxygen.

Carbon Dioxide. Carbon dioxide (CO_2), a minor component of the atmosphere, is highly soluble in water. Most carbon dioxide in pond water occurs as a result of respiration. Levels usually fluctuate inversely to dissolved oxygen, being low during the day and increasing at night, or whenever respiration occurs at a greater rate than photosynthesis. Carbon dioxide is present in surface water at less than 5 mg/l (5 ppm) concentrations, but may exceed 60 mg/l (60 ppm) in many well waters and 10 mg/l where fish are maintained in large numbers. Some aquatic animals, including fish, survive at up to 60 mg/l for short periods. If oxygen is lowered into its stress-causing range, carbon dioxide limitation is reduced to 20 mg/l.

Carbon dioxide interferes with the ability of the aquatic animal to extract oxygen from water, contributing to stress of fish during periods of low oxygen. Aerating water to improve its oxygen content drives off excess carbon dioxide.

Oxygen and carbon dioxide may be supplied to a body of water from the air. Oxygen and carbon dioxide may also be supplied when living things within the water carry on photosynthesis and respiration.

Ammonia. Test kits for determining ammonia in water measure total ammonia. To determine if a large percentage of the ammonia is in un-ionized form, pH is also measured. A pH above 8, in the presence of ammonia concentrations above 0.5 mg/l, is cause for concern.

Algae use ammonia as a nitrogen source for making proteins. Concentrations usually remain low in ponds with phytoplankton blooms. The greatest concentration of ammonia often occurs after plankton die-offs at which time pH is low due to high levels of carbon dioxide, and the majority of ammonia is present in the relatively nontoxic ionized form.

Nitrite (NO_2^-) is one of the basic products of organic matter decomposition. It acts as an intermediate stage in the conversion of ammonia to nitrate. These reactions occur in soils, mud, and water. Nitrite changes quickly to nitrate if oxygen is present. In culture ponds that are rich from feeding, a temporary accumulation of this chemical in harmful amounts sometimes occurs. Nitrite measurements of more than 1 mg/l (1 ppm) should be suspect in causing fish deaths.

The binding of nitrite with the hemoglobin molecule gives blood a chocolate brown color. Fish farmers call this condition *brown blood disease.* In medical terms, it is known as *methemoglobinemia.* The toxicity of nitrite to fish is lessened by the presence of chlorides in water. Most warmwater fish can tolerate at least 0.4 mg/l of nitrite in fresh water without treatment if oxygen levels remain above 4 mg/l.

Nitrite should be monitored frequently if a problem is suspected because its concentration may increase rapidly in pond water, especially during spring and fall, or when algae blooms suddenly die.

Nitrate (NO_3^-) is generally nontoxic to fishes and can be expected to occur at less than 2 mg/l in natural surface water. Fish can tolerate several hundred mg/l.

Total Alkalinity and Total Hardness

Total alkalinity and total hardness are measures of the basic substances of water. Because in natural water these substances are usually carbonates and bicarbonates, the measurement is expressed as mg/l of equivalent calcium carbonate. In special cases such as many groundwater and western ponds, sodium carbonate is the predominant basic substance. These basic substances resist change in pH (buffering) and where an abundance of calcium and magnesium bicarbonate are dissolved, the pH will stabilize between 8 and 9. If an abundance of sodium carbonate is present, the pH may exceed 9 or some laboratory forms report carbonate (CO_3^{2-}) and bicarbonate (HCO_3^-) in addition to total alkalinity. These are typically derived from alkalinity measurements by multiplication of standard conversion factors.

Total hardness is the measure of the total concentration of primarily calcium and magnesium expressed in milligrams per liter (ppm) of equivalent calcium carbonate ($CaCO_3$). Calcium and magnesium are usually present in association with carbonate as calcium carbonate or magnesium carbonate. Total hardness relates to total alkalinity and indicates the water's potential for stabilizing pH.

Fish do best when the measurement of hardness or alkalinity is between 20 and 300 mg/l. Below 20 mg/l can result in poor production. Cases where total hardness is considerably below the measure for total alkalinity are not desirable. Liming increases total hardness and total alkalinity.

TEST METHODS

Whether the water is tested in a lab or in the field by an aquaculturalist, general test methods include titrimetric, colorimetric, and electronic meters. For many water quality tests, companies sell test kits complete with all the chemicals and standards.

Titrimetric

Titrimetric analyses use a solution of known strength—the titrant—which is added to a known or specific volume of sample in the presence of an indicator. The indicator produces a color change indicating that the titration is complete. To calculate the results, the amount of titrant used is measured. A microburette or precision pipet adds the titrant.

Colorimetric

Beer's law states that the higher the concentration of a substance, the darker the color produced in a test reaction. This law provides the basis for determining the concentration of many substances in water samples. Known chemical reactions produce typical colors. The concentration that these colors represent is determined visually by comparing the color obtained from a sample to a set of standards. Because visual interpretation can be quite subjective, electronic colorimeters provide a more accurate indication of the color intensity. Colorimeters consist of a light source passing through a sample that is measured on a photo detector providing an analog or digital readout. Electronic colorimetric readings also compare the sample value to readings from a set of standard (known) readings.

Electronic Meters

Modern electronics provides aquaculturalists with a variety of electronic meters designed to measure specific water-quality factors, including pH, total dissolved solids, conductivity, dissolved oxygen, temperature, and turbidity. Like the chemical methods, standard solutions are important. They are used to calibrate the electronic meter.

Using test kits or electronic meters, aquaculturalists regularly check the oxygen, pH, carbon dioxide, and ammonia of water in production. Table 12-1 summarizes the test methods commonly used for aquaculture.

MATERIALS

Depending on the instructor, the materials for some of these measurements can differ. In most cases the instructor will use test kits that can be purchased from one of the suppliers listed. If test kits are not used for the dissolved oxygen and carbon dioxide, this exercise provides details of the reagents necessary. Also, test strips are available for the determination of pH, nitrites, and hardness. These may be used at the discretion of the instructor.

Each section of this lab requires two samples of water. These samples should be taken from sources that are likely to show different results. Some possible sources for water samples are:

➤ Pond water versus stream water

➤ Tap water versus pond or stream water

➤ Aquarium water versus tap water

➤ Water from a goldfish bowl recently cleaned and replenished with fresh water versus water from a goldfish bowl where the goldfish have fed, respired, and excreted for several days

For pH

➤ Water samples from two different sources

➤ Two small containers (flasks or beakers)

➤ Labels

➤ Either a test kit from HACH or LaMotte, AquaChek® test strips from Environmental Test Systems, Inc., or a pH meter

For Dissolved Oxygen

➤ Water samples from two different sources

➤ Two small containers (flasks or beakers)

➤ Labels

➤ Either a test kit from HACH or LaMotte, an oxygen meter, or all of the following:

   —Droppers (one for each solution)

   —Solution A—48 percent manganous sulfate

   —Solution B—70 percent potassium hydroxide and 15 percent potassium iodide

   —Solution C—concentrated sulfuric acid

   —Solution D—2 percent starch

   —Solution E—0.31 percent sodium thiosulfate

TABLE 12-1 SUMMARY OF WATER TESTING METHODS		
Parameter	**Procedure**	**Method**
Alkalinity	Collect samples in plastic or glass bottles. Fill completely and cap tightly. Avoid excessive agitation and prolonged exposure to air. Keep samples cool in refrigeration unit or ice chest, and analyze within 24 hours. Warm to room temperature before analyzing.	Test kit; titrimetric; test strips.
Ammonia	Collect samples in clean glass or plastic bottles. Samples not analyzed immediately may be preserved by reducing the pH to 2 or less with sulfuric acid. Refrigerate samples and analyze within 24 hours. Warm to room temperature and neutralize before analysis.	Test kit; colorimetric; visual or electronic colorimeter.
Carbon Dioxide	Collect samples in clean glass or plastic bottles. Fill completely and cap tightly. Avoid excessive agitation or prolonged exposure to air. Analyze as soon as possible but samples can be stored at least 24 hours by cooling to a temperature lower than the source.	Test kits; titrimetric.
Chloride—see Salinity		
Dissolved Oxygen—see Oxygen, Dissolved		
Hardness	Collect samples in plastic or glass bottles that have been washed with detergent, rinsed with tap water, and rinsed with 1:1 nitric acid solution and deionized water. Store only if prompt analysis is not possible.	Test kits; titrimetric and colorimetric, electronic.
Nitrate	Collect samples in clean plastic or glass bottles. Store at 39°F (4°C) if sample analyzed in 24 to 48 hours. Warm to room temperature before running test.	Test kits, colorimetric, visual and electronic. Electronic meter.
Oxygen, Dissolved	Measure directly in water source or collect samples. Sampling and sample handling important for meaningful results. Dissolved oxygen changes due to many variables.	Test kit; Winkler titration. Electronic meter.
pH, Water	Take readings at 10 in (25 cm) below the water surface. Collect samples in clean plastic or glass bottles. Fill bottles completely and cap tightly. If a probe is used, calibrate using a precision thermometer. Calibrate meter with standard buffers at pH 7 and pH 10.	Test kit; colorimetric; visual method. Calibrated electronic meter.
Salinity	Collect sample pooled from three levels if collected from a pond. Use clean glass or plastic bottles. Samples can be stored for up to 28 days in sealed containers.	Test kit; titration Test kit; electronic colorimeter. Meter; conductivity Electronic hydrometer; refractometer.
Secchi Disk Visibility—see Visibility, Secchi Disk		
Temperature, Water	Take readings at 10 in. (25 cm) below the water surface. Ideally, in a pond take readings also at mid-water, and 10 in. (25 cm) above the bottom. If probe is used, calibrate using a precision thermometer.	Certified thermometer, Electronic meter.
Total Dissolved Solids (TDS)	Collect samples, measure volume, dry, and weigh the dried sample. A convenient alternative is to test the conductivity, which can be used to estimate the TDS.	Gravimetric Electronic meter.
Visibility, Secchi Disk	At two locations in each pond, calculate Secchi Disk Visibility.	Secchi disk.

For Carbon Dioxide
➤ Water samples from two different sources

➤ Two small containers (flasks or beakers)

➤ Labels

➤ Either a test kit from HACH or LaMotte, or all of the following:

—Dropper

—Phenolphthalein solution

—0.4 percent sodium hydroxide solution

For Ammonia and Nitrite
➤ Water samples from two different sources

➤ Two small containers (flasks or beakers)

➤ Labels

➤ Either a test kit from HACH or LaMotte, or AquaChek® test strips from Environmental Test Systems, Inc.

For Hardness
➤ Water samples from two different sources

➤ Two small containers (flasks or beakers)

➤ Labels

➤ Either a test kit from HACH or LaMotte, or AquaChek® test strips from Environmental Test Systems, Inc.

CAUTION: All chemicals used are harmful to skin and clothing. If you spill any chemical, rinse with water and call your teacher immediately. Read and follow material safety data sheets that are included with all chemicals.

PROCEDURES

Depending on the resources of each school, the procedures will vary for determining pH, dissolved oxygen, carbon dioxide, ammonia, nitrite, and hardness. Throughout the procedures, use the same two water samples. Record the temperature of these samples in Table 12-2.

Determining the pH
1. Obtain 100 ml of two different water samples.

2. Follow directions for using test kits from HACH or LaMotte or test strips from Environmental Test Systems, Inc.

3. If available, use a pH meter.

4. Record results in Table 12-2.

TABLE 12-2 RESULTS OF WATER TESTING

Test	Unit	Sample 1 Source _____	Sample 2 Source _____
pH			
Dissolved oxygen			
Carbon dioxide			
Ammonia			
Nitrite			
Hardness			
Temperature			

Testing for Dissolved Oxygen

1. Obtain 100 ml of two different water samples.

2. Place the samples in small containers (flasks or beakers). If you are to pour samples into the flasks from large containers, pour slowly to avoid bubbling (aerating) the water. If you are collecting samples directly from the source, open the flasks under the water so that the flasks fill with water below the surface.

3. Label the containers with the source of the water. Also record in Table 12-2 where each sample was obtained.

4. If a test kit from HACH or LaMotte is being used, follow the directions inside the kit. Otherwise, use the reagents in the following steps.

5. With a dropper, add ten drops of solution A to each water sample. Hold the dropper close to the water surface to avoid splashing.

6. With another dropper, add ten drops of solution B to each water sample.

7. Gently mix the contents by swirling the containers. Be careful to avoid forming bubbles.

8. After swirling, let the flasks stand for one minute. (If you are using seawater as a sample, it must stand for 15 minutes.)

9. With a third dropper, add 15 drops of solution C to each water sample.

10. Gently mix the contents by swirling the container. Again, be careful to avoid splashing.

11. While gently swirling each container, add five drops of solution D to the sample. A deep blue color will appear.

12. With a dropper, add solution E one drop at a time to each sample. The number of drops of solution E must be counted.

13. Add drops of solution E until the water sample becomes colorless. Swirl the water samples after the addition of each drop in order to determine the true color of the solution.

14. The amount of dissolved oxygen in water usually is described in terms of parts of oxygen per million parts of water. Convert the drops of solution E to ppm of oxygen in each container by dividing the numbers of drops of solution E by 20. Carry your divisions to one decimal place.

15. Record the ppm of dissolved oxygen for each sample tested in Table 12-2.

Testing for Carbon Dioxide

1. Obtain 100 ml of two different water samples. Place the samples in small containers (flasks or beakers).

2. Label the container with the water source. Also record the water source in Table 12-2.

3. If a test kit from HACH or LaMotte is being used, follow the directions inside the kit. Otherwise use the reagents in the following steps.

4. With a dropper, add five drops of phenolphthalein solution to each sample. Mix by gently swirling your flasks.

   Note: If a light pink color forms and stays, no carbon dioxide is present.

5. With a clean dropper, add sodium hydroxide one drop at a time to each sample. The number of drops of this chemical that you add must be counted. Swirl the water samples after the addition of every few drops to determine the true color of the water.

6. Add drops of sodium hydroxide until the water sample becomes light pink and remains pink after swirling.

7. Record the number of drops of sodium hydroxide needed to change the color of the sample.

8. Convert the drops of sodium hydroxide to ppm of carbon dioxide in each container by multiplying the number of drops used by five.

9. Record the ppm of carbon dioxide in each sample tested in Table 12-2.

Testing for Ammonia, Nitrite, and Hardness

1. Obtain 100 ml of two different water samples. Place the samples in small containers (flasks or beakers).

2. Label the container with the water source. Also record the water source in Table 12-2.

3. Using test kits from HACH or LaMotte, follow the instructions in the kit and record the results in Table 12-2.

4. Nitrite and hardness can also be determined using the AquaChek® test strips from Environmental Test Systems, Inc. Record these results in Table 12-2.

ANALYSIS

1. Explain the difference or similarity in the pH of the two water samples.

2. What processes increase the pH of water, and what processes decrease the pH of water?

3. Which water sample contains more dissolved oxygen?

4. Why is oxygen important to organisms living in water?

5. Which type of organism could provide oxygen to water?

6. What name is given to the process in which oxygen is produced or given off?

7. Considering that the process that gives off oxygen requires light, how might this process be influenced by changes in water depth?

8. Which water sample contains more carbon dioxide?

9. Carbon dioxide and light are both required for photosynthesis. If a shallow pond with many plants growing in it was used as a water source for testing, how might the amount of carbon dioxide in the pond vary from day to night? Explain.

10. Which sample contains the highest level of ammonia and nitrite? Why?

11. What are some of the potential sources of ammonia and nitrite in a water sample?

12. Which water sample is the hardest? Why?

13. What makes water hard?

14. Why does a scientist not rely on only one trial when performing an experiment?

15. What effect does temperature have on dissolved oxygen, ammonia, and pH?

SUPPLIERS

Environmental Test Systems, Inc.
P.O. Box 4659
Elkhart, IN 46514
(http://www.aquachek.com)

HACH Company
P.O. Box 389
Loveland, CO 80539
(http://www.hach.com)

LaMotte Company
P.O. Box 329
Chestertown, MD 21620
(http://www.lamotte.com)

LAB 13 — Constructing a Secchi Disk to Measure Water Turbidity

INTRODUCTION

A Secchi disk is used to evaluate water turbidity, an indicator of water quality. Secchi disks can be purchased from supply companies or constructed of sheet metal, plexiglass, or masonite. In this lab, students will construct a Secchi disk.

The instructions that follow are for a sheet-metal Secchi disk with an 8-in. diameter. Depending on the equipment available at a school, the Secchi disk can be constructed from other materials.

CORRELATION

This lab and/or Lab 12 can be used with Chapter 12 of *Aquaculture Science*, 3rd Edition

BACKGROUND

Turbidity is caused by suspended solid matter that scatters the light passing through the water. Turbidity may be caused by sediment from disturbed or eroded soil clouding the water. Plankton also contributes to high turbidity when their numbers increase due to excess nutrients and sunlight.

When suspended sediment increases turbidity, this blocks out light needed by aquatic vegetation and bottom-dwelling (benthic) creatures. Suspended particles can also carry nutrients and pesticides through the water, and particles near the surface may absorb additional heat from sunlight, raising the surface water temperature.

In lakes, ponds, rivers, and estuaries, the Secchi disk is often used to measure turbidity. Secchi depth measurements are quite repeatable. The Secchi disk is painted with two black and two white quadrants (see Figure 13-1). The disk is attached to a calibrated line and lowered into the water until it just disappears from sight. This depth is recorded and the process is repeated.

Algae concentration can be estimated by placing a Secchi disk in the water to measure turbidity. In general, visibility through pond water should be no less than 18 to 20 in. from the water's surface. Secchi disk readings of 10 to 14 in. and a water coloration of yellowish-brown confirm adequate preparations and optimal conditions for best stocking results. A productive pond should have a good plankton density. A Secchi disk reading of about 12 in. should be maintained. For example, baitfish broodstock ponds should be fertilized to ensure that phytoplankton production is sustained all summer. Secchi disk readings should be 18 in. or less.

When phytoplankton is the principal source of turbidity, the Secchi disk depth is a measure of phytoplankton density. When phytoplankton is excessive, the only oxygen produced is near the water surface due to the light-screening effect of the phytoplankton. Generally, enough oxygen to support fish exists at depths two to three times the Secchi disk depth. Risks of a dissolved oxygen deficiency are greatly reduced when Secchi disk readings are in the 12- to 24-in. range.

FIGURE 13-1 The Secchi disk is lowered into water until it just disappears from sight.

MATERIALS

➤ 12-in. × 12-in. piece of sheet metal

➤ Sheet-metal snips

➤ Eye bolt

➤ Lead weight with a hole in the center

➤ Compass

➤ Awl

➤ Flat marine paint that will adhere to sheet metal, one can of white and one can of black

➤ 5 ft. of nylon or plastic rope with calibration marks every inch

➤ Yardstick or tape measure

➤ Pen or pencil

PROCEDURES

In this lab, you will construct a Secchi disk and then use it to measure turbidity.

Construction

1. Use a compass to mark an 8-in. diameter circle on the square of sheet metal. A pencil attached to a string can be used in place of a compass.

2. Cut out circle evenly with sheet metal snips. Be careful not to bend the metal or cut your fingers.

3. Paint the top of the disk with flat white paint and allow to dry.

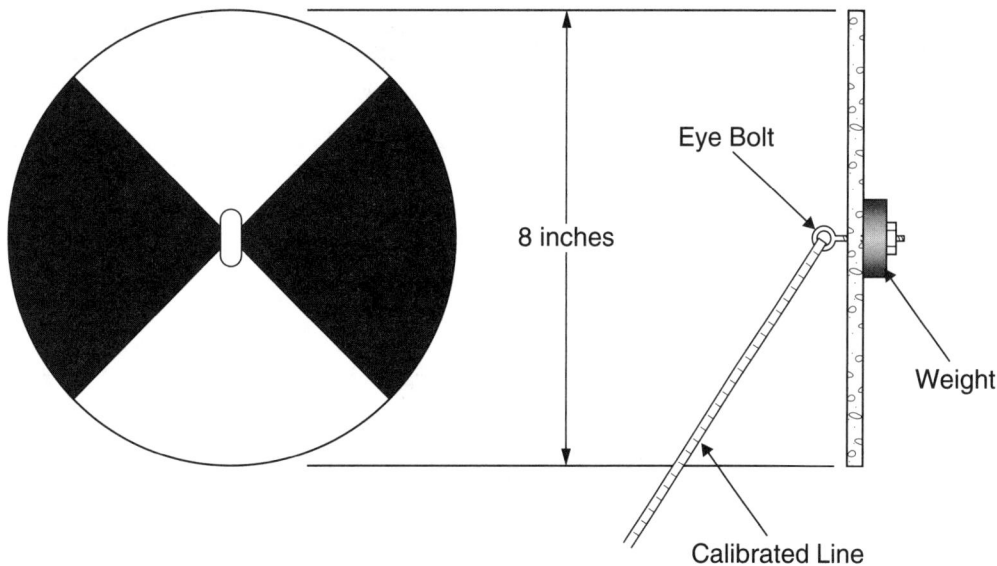

FIGURE 13-2 Dimensions and descriptions for constructing a Secchi disk.

4. Draw two perpendicular diameter lines on the white surface to create quadrants as shown in Figure 13-2.

5. Where the two lines cross is the center of the disk. Make a small hole in the center with an awl.

6. Paint two opposite quadrants flat black and allow to dry.

7. Attach lead weight and graduated line with eye bolt as shown in Figure 13-2. The weight will be attached to the underside of the disk to allow it to sink readily. A diving weight works well on disks of sheet metal, plexiglass, or masonite. A large, heavy magnet can be used on a disk made of sheet metal.

Measuring Turbidity

To standardize readings with a Secchi disk, measurements should be taken on clear, sunny, calm days after sunrise and before sunset.

1. Record the date, time, and location of the test in Table 13-1.

2. Lower disk into water until it just disappears.

3. Viewing the disk directly from above, read the calibrated line or yard (meter) stick and have your partner record this measurement in Table 13-1.

4. Lower the disk a little more, and then raise it until it just reappears.

5. Viewing the disk directly from above, read the calibrated line or yard (meter) stick and have your partner record this measurement in Table 13-1.

6. Add the two readings and divide by 2 to find the average, and record it in Table 13-1. This figure will be the Secchi disk visibility measurement.

7. Change the roles and repeat the test several times for comparison and practice.

TABLE 13-1 SECCHI DISK MEASUREMENTS						
Location	Date	Time	1st	2nd	Average	Notes

ANALYSIS

1. How do your Secchi measurements compare to the others?

2. What is responsible for the turbidity in your water source?

3. Why is the Secchi disk black and white?

4. Why would a large pond require seven Secchi readings in different locations to accurately determine the turbidity?

5. What difficulties did you experience when making your Secchi measurements?

LAB 14 — *Volume Calculations*

INTRODUCTION

Before any calculation is made of water volume, the units of measurement should be determined. The unit of measurement selected should be familiar and convenient for the specific situation. For example, the large volume of water in ponds is usually expressed as acre-feet, whereas the volume of a small tank may be expressed in gallons or cubic feet. Another decision is whether to use the English or metric system of measurement. A working knowledge of both systems is important because reports or publications may use either one. Metric units are easier to work with when small volumes or weights are involved.

In this lab, the volumes of different containers will be determined by measuring the containers and then calculating their volume in English and metric units.

CORRELATION

This lab can be used with Chapter 13 of *Aquaculture Science*, 3rd Edition.

BACKGROUND

The following formulas will be used to determine volumes of rectangular and cylindrical containers:

Volume of a cube = length × width × height *or* V = L × W × H

Volume of a cylinder = diameter squared × 0.7854 × length (height) *or*

$V = D^2 \times 0.7854 \times H$

Also, Table 14-1 provides the necessary conversion factors.

TABLE 14-1 CONVERSION FOR UNITS OF VOLUME								
	TO							
FROM	**ml**	**liter**	**m³**	**ft.³**	**fl. oz.**	**fl. pt.**	**fl. qt.**	**gal.**
ml	1	0.001	1×10^{-6}	3.53×10^{-5}	0.0338	0.00211	0.00106	2.64×10^{-4}
liter	1,000	1	0.001	0.0353	33.81	2.113	1.057	0.2642
m³	1×106	1,000	1	5.31	3.38×104	2,113	1,057	264.2
ft.³	2.83×10^4	28.32	0.0283	1	957.5	59.84	29.92	7.481
fl. oz.	29.57	0.0296	2.96×10^{-5}	0.00104	1	0.0625	0.0313	0.0078
fl. pt.	473.2	0.4732	4.73×10^{-4}	0.0167	16	1	0.5000	0.1250
fl. qt.	946.4	0.9463	9.46×10^{-4}	0.0334	32	2	1	0.2500
gal.	3,785	3.785	0.0038	0.1337	128	8	4	1

cm³ = cubic centimeter = milliliter = ml; m³ = cubic meter; ft.³ = cubic foot; fl. oz. = fluid ounce; fl. pt. = fluid pint; fl. qt. = fluid quart; gal. = gallon

MATERIALS

➤ Three cubes, rectangles or squares, of different sizes. These can be cardboard boxes or baking pans. Label these cubes A, B, and C (see Figure 14-1).

➤ Three cylindrical containers of different sizes. These can be plastic buckets. Label these containers D, E, and F (see Figure 14-1).

➤ Tape measure, ruler, or yardstick

➤ Pencil, paper, and calculator (optional)

PROCEDURES

1. Measure each of the containers and record this information in Table 14-2.

2. Calculate the cubic feet of water that each container will hold and add this information to Table 14-2.

3. Using the conversion factors in Table 14-1 convert the cubic feet answers to the other measures of volume indicated in Table 14-2.

4. Pick one of the cubic containers and one of the cylindrical containers and fill it with water. Measure the actual volume of water that the container will hold and compare this with the calculated volume.

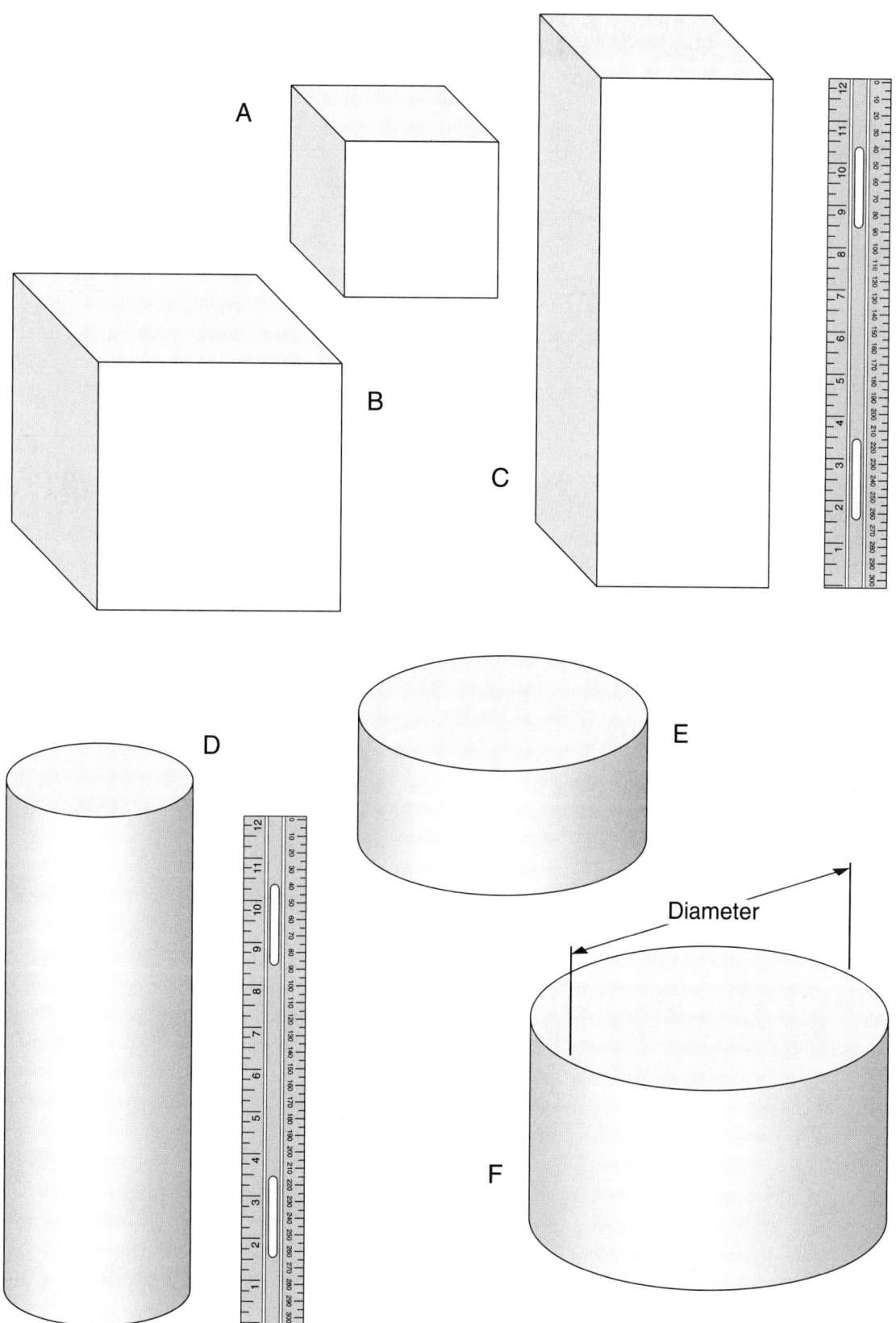

FIGURE 14-1 Common shapes used in aquaculture include cubes and cylinders.

TABLE 14-2 MEASURING AND CALCULATING WATER VOLUMES							
Container	Length	Width	Height	Ft.³	Gal.	Liters	m³
A							
B							
C							
	Diameter						
D							
E							
F							

ANALYSIS

1. Which measure of volume is the best for an aquaculturalist to use?

2. Why do aquaculturalists need to know how to calculate water volumes?

3. How accurate is the actual measured volume of water as compared to the calculated volume?

LAB 15 — *Aquarium Projects*

INTRODUCTION

An aquarium functions like a small pond. It can teach some principles of aquaculture. The same set of rules and laws that control a big fish farm can be seen in a small aquarium, right in your own room. Supplies for them can be found almost anywhere in the United States. Also, tropical aquarium fish come in over 1,000 varieties, so one can find a fish to suit almost any need.

The purpose of this lab is to conduct a fish-raising project in an aquarium to use and experience the knowledge needed on a larger scale for aquaculture.

CORRELATION

This lab can be used with Chapter 14 of *Aquaculture Science*, 3rd Edition.

BACKGROUND

The object when establishing an aquarium is to create a small pond. A pond is not a sterile place, and the aquarium should not be sterile either. Many aquarium manuals will tell the reader to bleach their gravel, tank, and everything else, sometimes on a regular schedule, making it a sterile place. Though this will kill many of the bacteria that could cause problems in an aquarium, it will also kill the helpful ones. Aquaculturalists learn about many interactions in ponds and water. If most of the ecosystem is killed with bleach or something else, it cannot be examined thoroughly. A good, healthy aquarium with nutrients, plants, and bacteria allows the study of the rules and laws that control aquaculture.

SAMPLE AQUARIUMS AND PROJECTS

Following are two suggested projects, one for the beginner and one for the more advanced. Procedures can be followed exactly, but they are designed to be guides, not rules. The vast variety of fish, plants, and aquariums will not allow for all the possibilities to be described. Once the principles are learned, some creative projects can be tried. The only rule is to keep the aquarium as natural as possible to learn about aquaculture.

Project #1

The purpose of this project is to produce some live-bearing fish and to keep some records similar to those of a commercial fish farmer. Livebearers are extremely domesticated and do well in an aquarium. Because the young come out of the female ready to swim and eat, they are easier to raise for the beginner. With enough cover for the young to hide in, only one tank is needed for both the parents and the young.

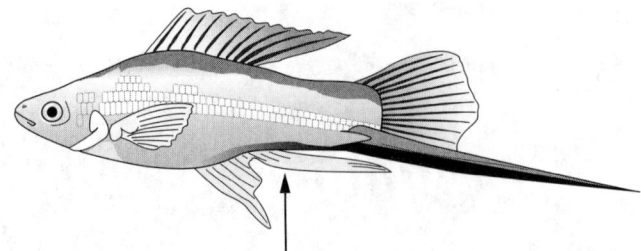

FIGURE 15-1 Specially-shaped fin on the underside of a male live-bearer fish.

The male of the livebearers has a specially shaped fin on his underside that he uses to fertilize the eggs inside the female (see Figure 15-1). This is one way to tell the males from the females.

The male chases the females around the tank. Then getting next to them, he will bend this special fin around so that he can mate. After this, the eggs will start developing inside the female, and it will take another few weeks before she has her young. As the young mature, a dark spot develops at the bottom of the female's belly. In some fish, even the eyes of the young can be seen as they grow.

As soon as the young hatch, they can swim and eat small pieces of food. Crumbling the flake food into a powder makes a good feed for the young. At each feeding, the adults should be fed first with regular flake food. With good plant cover for the young to hide in and regular feedings, many should survive. The parents may eat some of the young, and this is quite natural.

Feeding the Tank

Most of the mistakes in aquaculture are made in feeding, both by underfeeding and overfeeding. Underfed fish will become sickly and more apt to catch diseases and certainly will not reproduce well. Overfeeding is a more common problem. This causes poor water quality, putting the fish at risk of disease.

Feeding fish is easy to do, using a simple rule of thumb that fish farmers follow. Feed fish once or twice a day, the amount they eat in about ten minutes. If they are still hungry, they need more food. If they do not eat all the food, they received too much food. Following this rule of thumb will eliminate most of the problems in aquaculture. When fish are overfed, any food lying on the bottom of the tank should be removed. This is done by using a 5- to 6-ft. long piece of hose to siphon the uneaten food into a bucket.

Once the amount of food eaten is established, it should be recorded. This is important information because if fish ever stop eating as much, it may mean problems. It is a good indication that something is wrong with the fish or the water.

Cleaning the Aquarium

Even an aquarium that is well balanced with fish, plants, and light requires some changing of the water. The easiest way to do this is by using a siphon hose, vacuuming any wastes off the bottom at the same time. When adding new water:

1. Remove any chlorine, either by letting the water stand for 24 hours before adding to the aquarium or by using a dechlorinator. Most cities put enough chlorine into the drinking water to kill fish. In addition, some cities have ammonia in their tap water.

2. Add water that is the same temperature as the water in the aquarium. If the water is colder, add it slowly, allowing the fish to get used to it.

If the fish are not overfed, the plants and the filter should take care of the wastes in the aquarium. If the water turns yellowish or green, changing one-quarter to one-half of the water will usually fix the problem. A well-balanced aquarium should never need to be totally emptied.

A certain amount of algae will grow on the glass of the aquarium. Algae should be cleaned off the inside of the front and sides of the aquarium, but it can be left on the back of the tank. Many commercial scrapers are specifically made for this, but a plastic scrub pad will work. A scrub pad used around the house to clean dishes or anything else should not be used.

The filter also requires some maintenance. Outside power filters are convenient to clean. Most are sold with cartridges that just slip out for cleaning or replacement. To clean the cartridge, rinse it gently under a stream of cool water. The bacteria that help to break down the wastes in the aquarium live in this cartridge. A hot stream of water will destroy these bacteria. If the cartridge is too dirty to clean, replacement is necessary.

Broodfish in a properly maintained 10-gal. aquarium will start filling the aquarium with young. At this point:

1. Start selling some of the young or give them to a friend.

2. Set up another aquarium where you can raise them.

In an aquaculture operation, it is normal to have at least two separate tanks, one for the breeders and one for raising the young.

To get a good idea of the problems and solutions of breeding fish, this project needs to be continued for at least six months. Whenever anything is done with the aquarium, write it down. Table 15-1 is a log that should be photocopied and used for the duration of this project.

Project #2

This project is designed to use a system that is more common in commercial operations using two separate aquariums—the breeding tank and the rearing tank. The size of the aquariums needed will depend on the type of fish used for the project. Some species can reproduce in 1-gal. bowls, and others require at least a 40-gal. aquarium. This advanced project requires some research to determine the best kind of fish to raise. Libraries and pet stores provide information needed before deciding on which fish to use.

The equipment and methods needed to establish the aquariums will be the same as for the beginners' project, with some specialized additions for a few fish types. Some fish need caves, mops, slates, and other places for their spawning to take place.

Possibly, the main addition to this project is the need to produce live foods for the fry. The easiest way to do this is by hatching brine shrimp eggs, which can be found at most aquarium dealers and come with hatching instructions. Following is a list of equipment needed and directions on how to hatch them.

Describe your project in detail including any problems.

Materials
- ➤ Non-iodized salt (rock salt)
- ➤ Small aquarium air pump
- ➤ Air line tubing and air stone
- ➤ Brine shrimp net, or fine cloth filter
- ➤ 1-gal. jar

TABLE 15-1 AQUARIUM LOG

Beginning Date _____   Ending Date _____

Type of Fish Used _____

Number of Males _____   Number of Females _____

Size of Aquarium(s) _____ gallons

_____

_____

_____

Production Data

Date	No. of Young	Deaths	Feeding Notes

Costs

Aqu	Fish	Plants	Filter	Light	Food	Stand	Other

Directions

1. Soak the eggs for one hour in fresh water.

2. Add the salt according to the directions on the package of eggs. Different eggs will have different requirements.

3. Aerate the water in the jar.

4. After 24 hours, turn off the aeration. The baby brine shrimp will sink to the bottom, and the unhatched eggs and shells from those that hatched will float to the top.

5. Using a siphon, take the baby brine shrimp out of the bottom, filtering them through a net. Try to avoid stirring the jar's contents and avoid siphoning the shells with the baby brine shrimp.

6. Feed the baby brine shrimp to fish fry, making sure not to add too many. Brine shrimp will die in freshwater, and like fish food, will spoil the water.

When excesses are hatched, they can be stored in the refrigerator for several days in about 1-in of salt water from the hatching jar. Lab 5, Culturing Shrimp, provided practice on culturing brine shrimp.

This advanced project requires the same records as the beginners' project, with more information and more detailed observations.

Water Quality

Establishing an aquarium also provides an excellent opportunity to monitor water quality using the tests described in Lab 12, Testing Water: pH, Dissolved Oxygen, Carbon Dioxide, Ammonia, Nitrite, and Hardness. Use Table 15-2 to record the measurement of these tests on an aquarium over a period of time.

ANALYSIS

1. After keeping records on the aquarium for three to six months, use these records to write a report on the successes and failures of the project. Include the following items:
   - Total expenses
   - Type and amount of food
   - Any income from fish sales
   - List of problems and solutions
   - Sources of information

2. Using data from the records on water quality of the aquarium (Table 15-2), plot this data. On graph paper or with a computer spreadsheet, plot the pH, dissolved oxygen, carbon dioxide, ammonia, and hardness by date.

TABLE 15-2 AQUARIUM WATER QUALITY					
Date	pH	Dissolved Oxygen	Carbon Dioxide	Ammonia	Hardness

(continued)

TABLE 15-2 AQUARIUM WATER QUALITY (*Continued*)					
Date	pH	Dissolved Oxygen	Carbon Dioxide	Ammonia	Hardness

LAB 16   Managing a Recirculating System

INTRODUCTION

A recirculating system is an enclosed system where the only water replacement is the water lost to evaporation and cleaning. Recirculation systems in aquaculture create a great deal of interest in the aquaculture community in the United States and other parts of the world. Given enough resources, most species of fish grown in ponds, floating net pens, or raceways could be reared in commercial scale recirculation systems.

Recirculating systems have generally been expensive to build and operate—increasing the cost of producing fish. Five major advantages in using recirculating aquaculture systems are:

1. Low water requirements

2. Low land requirements

3. Control of water temperature

4. Control of water quality

5. Independence from adverse weather conditions

In addition, they have great value as an educational lab. The aquaculturalist has the ability to measure and control most of the variables that make up the environment of the recirculation system. This makes the recirculation system a good tool in teaching knowledge and skills in agriculture, biology, and related sciences (see Figure 16-1).

Recirculating systems are ideal for educational labs. Students can learn to control and maintain the system to provide a favorable environment for the culture of fish or other aquaculture plants and animals. In the process of doing this, they can gain many competencies in agriculture, biology, physics, chemistry, mathematics, and agriculture. The purpose of the recirculating system should be education, not commercial production.

The purpose of this lab is to provide guidelines on the management of recirculation systems.

CORRELATION

This lab can be used with Chapter 15 of *Aquaculture Science*, 3rd Edition and will depend upon the access to a recirculating system.

FIGURE 16-1 Classroom recirculation system.

BACKGROUND

Recirculation systems are most often employed where sufficient water is not available to wash fish wastes out of the production tank. By recirculating water through a water treatment system that removes ammonia and other waste products, the same effect as a flow-through configuration is achieved. A key to successful recirculation production systems is the use of cost-effective water treatment systems.

All recirculation production systems use processes to remove solid wastes, oxidize ammonia and nitrite-nitrogen, and aerate and/or oxygenate the water (see Figure 16-2).

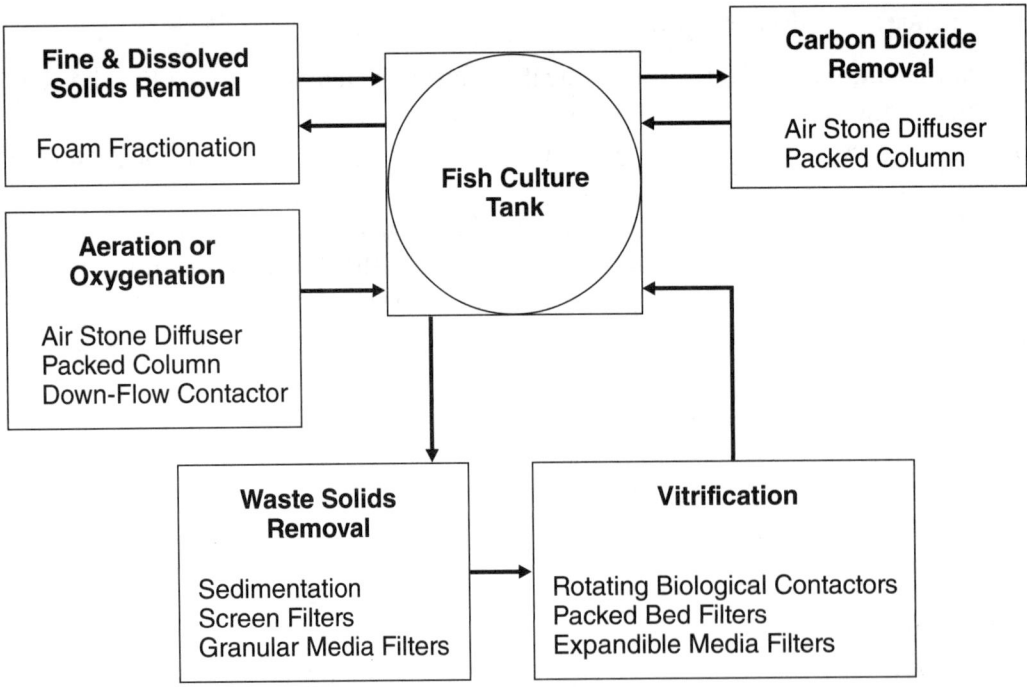

FIGURE 16-2 Required unit processes and typical components used in recirculating aquaculture production systems.

Waste Solids

The major components of feeds used in aquaculture production consist of protein, carbohydrates, fat, ash, and water. The portion of feed not used by the fish is excreted as an organic waste (fecal solids). Bacteria break down these fecal solids, along with uneaten feed, in the system, consuming oxygen and generating ammonia-nitrogen. To minimize their impact on water quality, waste solids need to be removed from the system as quickly as possible. Waste solids can be classified into four categories:

1. Settleable

2. Suspended

3. Floatable

4. Dissolved solids

In recirculation systems, the first two are of primary concern, while the other two can become problems in systems with very little water exchange.

Settleable Solids. Settleable solids are generally the easiest of the four categories to deal with and should be removed from the water in the tank as rapidly as possible. Settleable solids settle out of water within one hour under still conditions. Settleable solids can either be allowed to settle within round culture tanks (where they accumulate on the bottom in the center), or they can be kept in suspension with continuous agitation and removed with a properly designed sedimentation tank (clarifier) or filter. The sedimentation process can be enhanced through the addition of steeply inclined tubes (tube settlers) within the sedimentation tank to reduce flow turbulence and promote uniform flow distribution.

Suspended Solids. From a fish producer's point of view, the difference between suspended solids and settleable solids is a practical one. Suspended solids will not settle out of the water column under still conditions within one hour and would not be expected to be removed by conventional settling. If not removed, suspended solids can significantly limit the amount of fish that can be grown in the system and can interfere with and irritate the gills of fish.

The most popular treatment method for removing suspended solids generally involves some form of mechanical filtration. The two types of mechanical filtration most commonly used are screen filtration or granular media filtration (sand or pelleted media).

Fine suspended solids (less than 30 microns) contribute to more than 50 percent of the total suspended solids in a recirculation system. These solids increase the oxygen demand of the system and have been shown to cause gill irritation and damage in finfish. Additionally, dissolved organic solids (proteins) can contribute significantly to the oxygen demand of the total system.

Fine and dissolved solids cannot be easily removed by sedimentation or mechanical filtration technology. Foam fractionation (also referred to as *protein skimming*) has been widely used to remove these solids in recirculation tank systems. Foam fractionation, as used in aquaculture, introduces air bubbles at the bottom of a closed column of water that creates foam at the air/water interface. As the bubbles rise through the water column, fine suspended solid particles attach to the bubbles' surface, creating the foam at the top of the water column. The foam buildup is then channeled out of the fractionation unit to a waste collection tank. Efficiency of a foam fractionation system is dependent on the properties of the water in the system (salt concentration, temperature, pH, and so on), but can significantly reduce water turbidity and oxygen demand in the culture tank.

Nitrogen

Total ammonia-nitrogen (TAN) consists of two fractions:

➤ Un-ionized ammonia (NH_3)

➤ Ionized ammonia (NH_4^+)

TAN is the by-product of protein metabolism. It is excreted from the gills of fish as they assimilate feed and is produced when bacteria decompose organic waste solids within the aquaculture system. The un-ionized form of ammonia-nitrogen is extremely toxic to fish. The fraction of TAN in the un-ionized form is dependent upon the pH and temperature of the water (see Figure 16-3). At a pH of 7.0, most of the TAN is in the ionized form, whereas at a pH of 8.0 the majority is in the un-ionized form. Reduction in growth rates may be the most important sublethal effect. In general the concentration of un-ionized ammonia-nitrogen in tanks should not exceed 0.05 mg/l.

Nitrite-nitrogen (NO_2^-) is a product of the oxidation of ammonia-nitrogen. Vitrifying bacteria (*Nitosomonas*) in the production system use ammonia-nitrogen as an energy source for growth and produce nitrite-nitrogen as a by-product. These bacteria are the basis for biological filtration. The vitrifying bacteria grow on the surface of the biofilter substrate and to some extent on all production system components including pipes, valves, tank walls, and so on. Although nitrite-nitrogen is not as toxic as ammonia-nitrogen, it is harmful to aquatic species and must be removed from the system. Concentrations of nitrite-nitrogen should not exceed 0.5 mg/l for long periods of time. Fortunately, Nitrobacter bacteria that are also present in most biological filters use nitrite-nitrogen as an energy source and produce nitrate as a by-product. Nitrates are not generally of great concern to the aquaculturalist. Studies have shown that aquatic species can tolerate extremely high (greater than 100 mg/l) concentrations of nitrate-nitrogen in production systems.

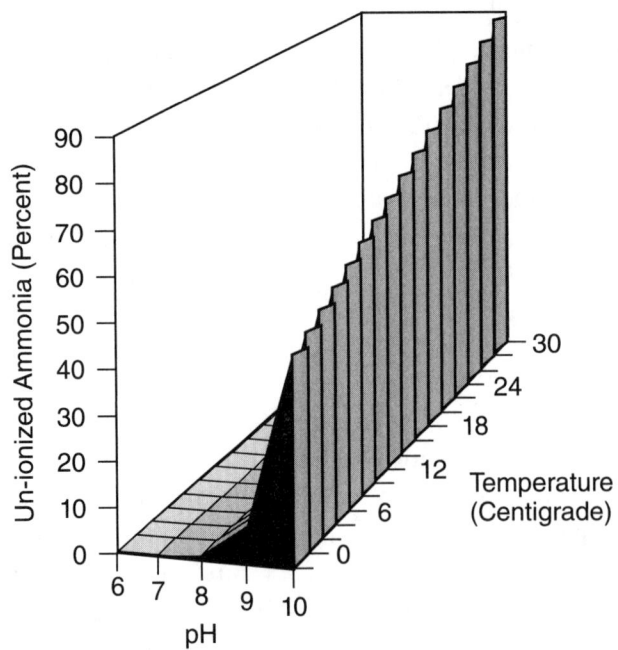

Effect of Increased Temperature and pH on Un-ionized Ammonia in Water

FIGURE 16-3 As temperature and pH increase, so does un-ionized ammonia.

Nitrate-nitrogen concentrations do not generally reach such high levels in recirculation systems. Nitrate-nitrogen is either flushed from a system during system maintenance operations (such as settled solids removal or filter backwashing) or denitrification occurs within a treatment system component such as a settling tank. Denitrification is mainly due to the metabolism of nitrate-nitrogen by anaerobic bacteria producing nitrogen gas that is released to the atmosphere during aeration processes.

Nitrogen Control. Control of the concentration of un-ionized ammonia nitrogen (NH_3) in the culture tank is the primary objective of recirculation treatment system design. Ammonia-nitrogen must be removed from the culture tank at a rate equal to the rate of production to maintain a safe concentration. A number of technologies available for removing ammonia-nitrogen from the water exist, including air stripping, ion exchange, and biological filtration. Biological filtration is the most widely used. In biological filtration (biofiltration), a substrate with a large surface area is provided for nitrifying bacteria's attachment and growth. Gravel, sand, plastic beads, plastic rings, and plastic plates are commonly used biofiltration substrates. The configuration of the substrate and its contact with wastewater define the water treatment characteristics of the biological filtration unit.

Fish

Many species are suitable for recirculation aquaculture, but tilapia are very tough fish and will survive when other species may die. So, if possible, they should be used as beginning species. However, hybrid striped bass do well, as do pure striped bass, largemouth bass trained on artificial feed, minnows, sunfish, and ornamentals, such as gold fish and guppies.

PROCEDURES

The design of a recirculation aquaculture system for use in an educational setting should recognize that experienced personnel may not operate the system. So, the system needs to be forgiving of operator error. To realize this in a design, each system component should be over designed for its function. In many cases, the system will need to operate as a stand-alone unit with little support from the existing physical plant. Additionally, the system should operate in a classroom or laboratory setting with a minimum of room modification and distractions.

Routine Maintenance

Very little routine maintenance needs to be performed on the fish tanks. The smooth bottomed, circular fish tanks provide a self-cleaning flow that directs most of the uneaten feed and fish feces to the center drain. Once a day, the screens on the tank drains should be removed and cleaned as necessary.

Every second or third day of operation, the solids removal basin will need cleaning. This operation will require that the submersible pump be unplugged and the tube settlers removed from the basin. The accumulated solids are siphoned from the bottom of the basin.

The coarse and fine polyester fiber screens should be removed and cleaned by spraying with a high-pressure hose. Upon replacing the vertical filters, you may restart the submersible pump. New water from the tap should be added to one of the fish tanks to replace the water removed in the solids siphoning operation (5 to 10 gallons). The only other water loss from the system will come from the container into which the foam fractionator discharges.

Once a week, or as needed, the drain lines should be cleaned. A large bottle brush (1½ in. diameter) should be attached to a long 12-foot stiff wire (electricians wire chase) and pulled or pushed through each straight section of the drain. This procedure will remove the slim buildup on the pipes, reducing the frictional losses. Failure to clean the drainage system will result in a lowering of the water level in the water treatment basins.

All other system components should be maintained according to the manufacturers' specifications and guidelines.

System Management

Management of a recirculation system can be satisfying, confusing, and depressing all at the same time. Management parameters and observations should be recorded in Tables 16-1 and 16-2.

TABLE 16-1 RECIRCULATION SYSTEM MONITORING									
Date	System I.D.	Average Total NH_3	Alkalinity	Chloride	Average Nitrite	Temp.	Average DO	CO_2	Observations/Comments

TABLE 16-2 FEED CONSUMPTION CHART FOR RECIRCULATION SYSTEM						
Date	Tank 1	Tank 2	Tank 3	Tank 4	Tank 5	Observations/Comments

Water Quality Monitoring. Water quality monitoring begins with observing the fish because their actions will often reveal many problems. Write down observations of day-to-day fish behavior along with daily recording of the test data on the system (Tables 16-1 and 16-2). A systematic monitoring of both the water quality and fish behavior will pay dividends at a later time.

The monitoring equipment used is a matter of preference and budget. A kit that will allow you to test oxygen, nitrite, carbon dioxide, alkalinity, ammonia, and chlorides is essential. Most kits contain some version of the Winkler method of testing oxygen, which is good but very time-consuming. Oxygen requires several tests per day, so if budgets permit, an oxygen meter is a good investment. An Imhoff cone is useful to determine the solids in your system, and a good thermometer will help monitor and maintain the proper water temperature for the fish.

When correcting a parameter, do so gradually as a rapid change may be more harmful in the long run if done too quickly. The key is to make the changes gradually giving both the biofilter and the cultured species a chance to adapt to the new conditions. The possible exception to this is low dissolved oxygen (DO). If this is very low, it should be raised as quickly as possible.

For details on monitoring water quality, refer to Lab 12.

Sunlight. Sunlight is not a water quality parameter, but it can affect the water quality. If the system is located near a window or in a greenhouse, some provision must be made to reduce the light because it will cause excessive algae growth, which in turn may cause large variations in pH, DO, CO_2, nitrate toxicity, ammonia toxicity, and biofilter fouling.

Excess algae also may lead to off flavors in the fish when harvest time rolls around.

SUMMARY

Managing a biofilter is related to all of the water quality tests. Keeping the biofilter functioning is the most important job in recirculation. A healthy biofilter has a very thin growth of orange-brown colored bacteria on its surface. Thick clumps of a brown, slimy material is likely to be heterotrophic bacteria and not vitrifying bacteria. These may lead to clogging of the filter and the pipes and should be discouraged by keeping the particles of uneaten feed and wastes out of the system as much as possible. Water quality tests must be looked at in two ways. What does it mean to the cultured species, and what does it mean to the biofilter?

People with more experience using recirculation systems are quite willing to share information and help where possible. Develop a network of experienced operators you can call when help is needed.

ANALYSIS

Write a lab report based on the notes in Tables 16-1 and 16-2. Respond to the remaining Analysis items in your lab report.

1. Discuss the cost of operation and cost of construction of your system.

2. Read Appendix F and indicate how these Do's and Don'ts apply to your system.

3. Identify the four categories of waste solids.

4. Describe the foam fraction and its removal.

5. What is the purpose of a biofilter and how can you tell when it is working in the system?

6. Discuss the features that make your system forgiving.

SUPPLIERS

Commercial, complete systems can be purchased rather than trying to construct a recirculation system. For example:

Carolina Biological Supply Company: http://www.carolina.com

(key words = aquaculture system)

Pond and Landscape Solutions: http://www.pondsolutions.com/recirculating-systems.htm

LAB 17 Aquaponics

INTRODUCTION

Aquaponics is the combination of aquaculture and hydroponics. Hydroponics is the growing of terrestrial (land) plants in nutrient-rich water without soil. Terrestrial plants normally grow with their roots anchored in the soil. The roots absorb nutrients from the soil. Hydroponics can be used in conjunction with artificial sunlight to extend the growing season. Also, many vegetables yield more per square foot in a shorter time than with soil. Nutrients are recirculated, resulting in less pollution.

Aquaponics reduces the need to filter fish wastes out of water and may eliminate or reduce the need to use fertilizer for plant growth. Water containing the fish wastes becomes the nutrient-rich water for the hydroponics.

In this lab, students will be able to design and use an aquaponics system appropriate for the type of aquaculture being demonstrated in the classroom. Designing an aquaponics system uses applications of physical and biological sciences.

CORRELATION

This lab can be used with Chapter 16 of *Aquaculture Science*, 3rd Edition.

BACKGROUND

The type of system designed will depend on the aquaculture system being used. Basically, three systems can be used:

1. Direct

2. Flow-through

3. Misting

In the direct system, the plants and the fish live together in the same water. Plant roots dangle in the nutrient-rich water and tend to filter the water as the nutrients are removed to support plant growth. Because the direct system is not entirely self-sustaining on a small scale, mechanical filters are used. This system is often used in classroom aquaria. The direct system is the closest to a natural ecosystem. Fish have a direct relationship with the plants. Obviously, plant-eating fish cannot be grown in a direct system.

The flow-through system uses pumps and gravity flow to move water to a location away from the fish tank for the plants. It requires a larger set up such as a recirculating system and a lab or outdoor area. Flow-through systems use PVC pipes, pumps, and heaters (see Figure 17-1). In a flow-through system, the fish and the plants are easier to harvest because they are separated.

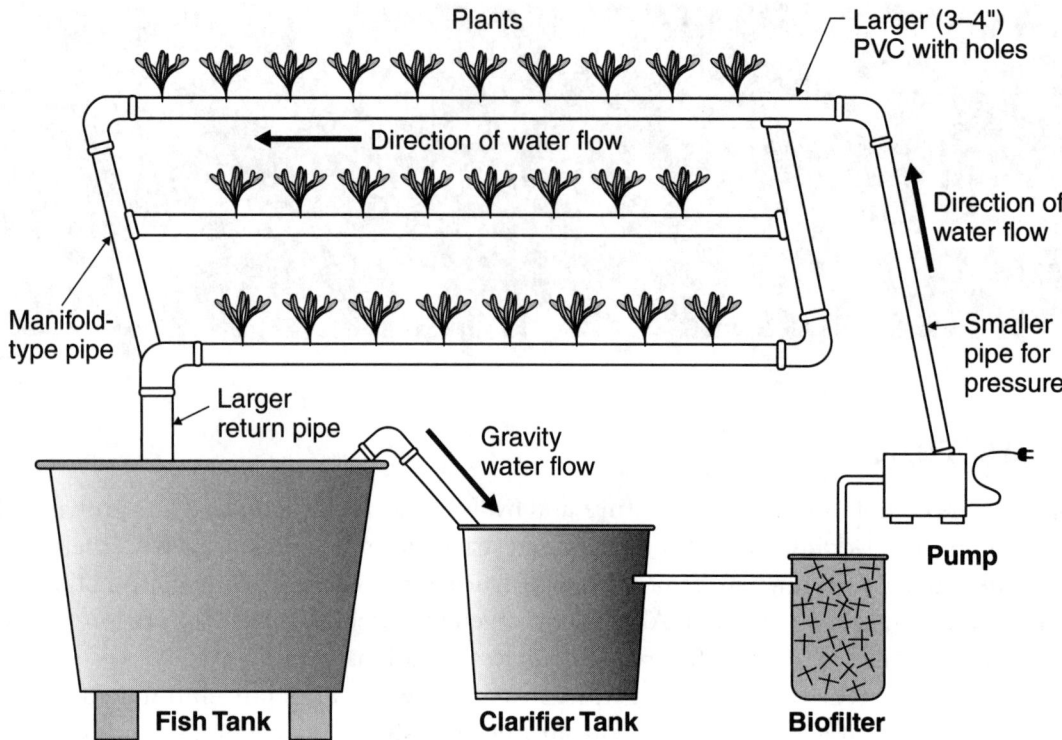

FIGURE 17-1 A simple, inexpensive flow-through system for a classroom.

Misting systems are found mainly in commercial production. This system involves spraying a fine mist of water onto the roots of the plants that are suspended in some way and not in a growing medium.

General Requirements

Besides a source of nutrient-rich water and the plants, aquaponics requires oxygen, pH monitoring, and an anchoring medium.

Oxygen. Oxygen is an important nutrient used by the plants' roots. In fact without ample oxygen to the roots, the plant will drown. Hydroponics is very generous to plant root systems allowing for easy exchange of oxygen and other vital gases.

pH (potential hydrogen). The pH level refers to the acid or alkaline level of the nutrient solution. The pH scale ranges from 0 to 14 with anything below 7 considered acidic. Most plants prefer a pH level of between 5.5 and 6.5. A pH level that is too high or too low can affect plants' ability to use nutrients.

The natural process of nitrification breaks down ammonia-nitrogen (produced by the fish as a metabolic waste product) into nitrite and then nitrate. This process also produces acids that lower the pH of the water. The optimum pH to maintain is 7.0. This level is a compromise between the optimum level for nitrification (7.5) and the optimum level for lettuce growth (6.5). To elevate the pH to this level, base should be added on a daily or every-other-day basis. The best bases to use in the system are potassium hydroxide (KOH) and calcium hydroxide (Ca(OH)$_2$). These are best because they do not increase the sodium salts in the system (which are toxic to the plants and can cause tip burn) and these bases provide essential nutrients to the plants. Potassium and calcium are required nutrient additions to ensure healthy plant growth.

Anchoring Medium. Growing media anchor the roots and give support to the plant. Because all nutrition necessary to grow plants is supplied by the nutrient solution, mediums anchor the plant. The most popular growing mediums in hydroponics are Heydite (a porous shale), Hydrocorn (clay pellets), Perlite (vermiculite), and Rockwool. Heydite is small pieces of shale rock that can be re-used indefinitely, making it environmentally friendly. Perlite, made from volcanic rock, is a white, lightweight material often used as a soil additive. The 1/8- to 1/4-inch pellets can be used alone as a growing medium, but they don't provide enough anchorage for large plants. Perlite is often used to start seed and cuttings, which can be easily transplanted after rooting. Vermiculite is used the same way as Perlite, and the two are sometimes mixed together. It is made from heat expanded mica and has a flaky, shiny appearance. Rockwool is also made of rock but has been melted, spun like cotton candy, and molded into growing blocks and slabs. Hydroponic media that work best are pH neutral, provide ample support for plants, retain moisture, and allow space for good air exchange.

Nutrients or Plant Food. Water containing fish wastes in the aquaculture system provides the nutrients. In a plain hydroponic system, the plant food is supplied to the plants by dissolving natural fertilizer salts in water to make a nutrient solution. Hydroponic growers mix their own nutrient solution or purchase a commercially prepared nutrient mix. If iron is deficient in the system, it is supplied by additions of iron chelate once every three weeks. The quantity supplied should keep the concentration of 2 ppm Fe in the system water.

MATERIALS

This lab provides details for a direct and a flow-through aquaponic system. The designs should be modified. Planning the system and determining the needed materials can be a class project.

Direct System

- ➤ Aquarium: Any watertight container with fairly vertical sides will work.

- ➤ Floating platform: Styrofoam 1½ to 2 inches thick.

- ➤ Plastic cups: Any 3-ounce brand with tapered sides.

- ➤ Growing medium: Enough to fill the plastic cups (Perlite or a Perlite/Vermiculite mix).

- ➤ Air pump and air stone.

- ➤ pH test kit.

- ➤ Nylon netting to protect plant roots from fish.

- ➤ Plants: lettuce, tomatoes, cucumbers (lettuce preferred).

Flow-Through System

- ➤ Plastic or fiberglass containers

- ➤ PVC pipe and/or flexible vinyl tubing

- ➤ Biofilter

- ➤ Pump

- ➤ pH test kit

➤ Plants

➤ Growing medium

➤ Fiberglass screen (depending on design)

➤ Gravel (depending on design)

➤ A plan

PROCEDURES: DIRECT SYSTEM

1. Light breaks down the nutrient solution and encourages algae growth, so the aquarium may need a light shield constructed of cardboard or aluminum foil to keep light out of the reservoir (aquarium). To view the roots, make the light shield (or part of it) removable.

2. Cut the Styrofoam float to fit the reservoir. Cut the float a little smaller than the opening so that it will not bind when the water level changes.

3. Cut the holes in the float to the proper size for the plastic cups. Holes should allow the bottoms of the cups to hang below the bottom of the float but not fall through. (Solo® brand 3-oz. plastic bathroom cups require a 17/8- to 2-inch hole.)

4. Cut several holes (approximately 1/8- to ¼-inch in diameter) in the bottom of the plastic cups.

5. Add growing medium to the cup. (**Note:** If the growing medium falls out through the holes, put a small piece of fiberglass window screen or small piece of cloth over the holes before adding the growing medium.)

6. Plant your seedling, rooted cutting, or seed in the growing medium.

7. Attach ¼-inch airline to the air stone and place air stone in reservoir. Attach free end of tubing to air pump and plug in air pump to outlet, make sure that there are bubbles coming from the air stone.

8. Place floating platform on top of the nutrient solution. Put plastic cups into the holes in the floating platform (see Figure 17-2).

9. Maintain a daily log of the construction and operation of the system (Table 17-1).

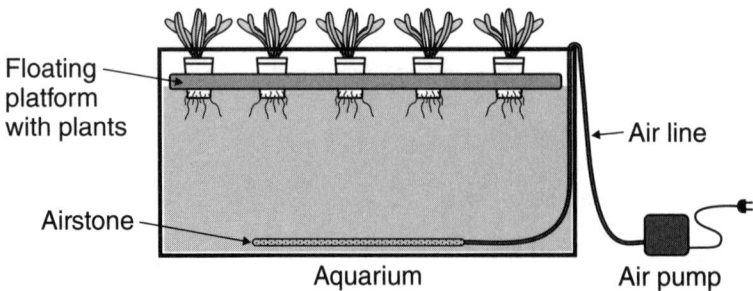

FIGURE 17-2 Design for a direct system using a classroom aquarium.

TABLE 17-1	TRACKING AN AQUAPONICS PROJECT		
Date	**Water Quality Observations**	**Plant Growth Observations**	**Other Observations**

PROCEDURES: FLOW-THROUGH SYSTEM

No matter what type of flow-through system is designed, the nutrient-rich water from the fish tank needs to be pumped or moved by gravity to a clarifier tank, through a biofilter, and then pumped or moved by gravity through the plant growing area. After the water flows through the plant growing area, it is returned to the fish tank. Many different designs and plans are available (see Figures 17-3 and 17-4).

Tanks used in the system should be glass, fiberglass, or plastic. In flow-through systems, PVC or flexible vinyl tubing connects the tanks and the plant growing areas. After all the connections are made, the system should be run for two weeks without any fish or plants.

A simple gravity system can be created using 5-gal. buckets (or equivalent), one filled with nutrient-rich water from the fish tank after clarification and another holding the growing medium and plants. Pumps can be used, or lifting the bucket containing the nutrient solution allows the nutrient solution to flow into the bucket containing the growing medium and plants to water the plants. To drain, simply lower the nutrient bucket and gravity drains the nutrient solution back into fish tank. Then follow these steps:

1. Drill holes in the clean plastic buckets on the side approximately ½ inch above the bottom of the bucket. (**Note:** The size of the holes will depend on the size of the tubing used. Tubing with ½ inch inside diameter requires a ½ inch hole.)

2. Insert tubing into the holes of both buckets approximately 2 in. Test the assembly for leaks by placing the buckets side-by-side and filling with water. (**Note:** The tubing should fit tightly so that no leaks occur. A leak from where the tubing connects to the bucket can be sealed from the inside of the bucket with some silicone sealant.)

3. Empty the water out of the assembly and place the gravel into the bottom of one of the buckets. This will be the planter. The other buckets will be the reservoir and clarifier tank.

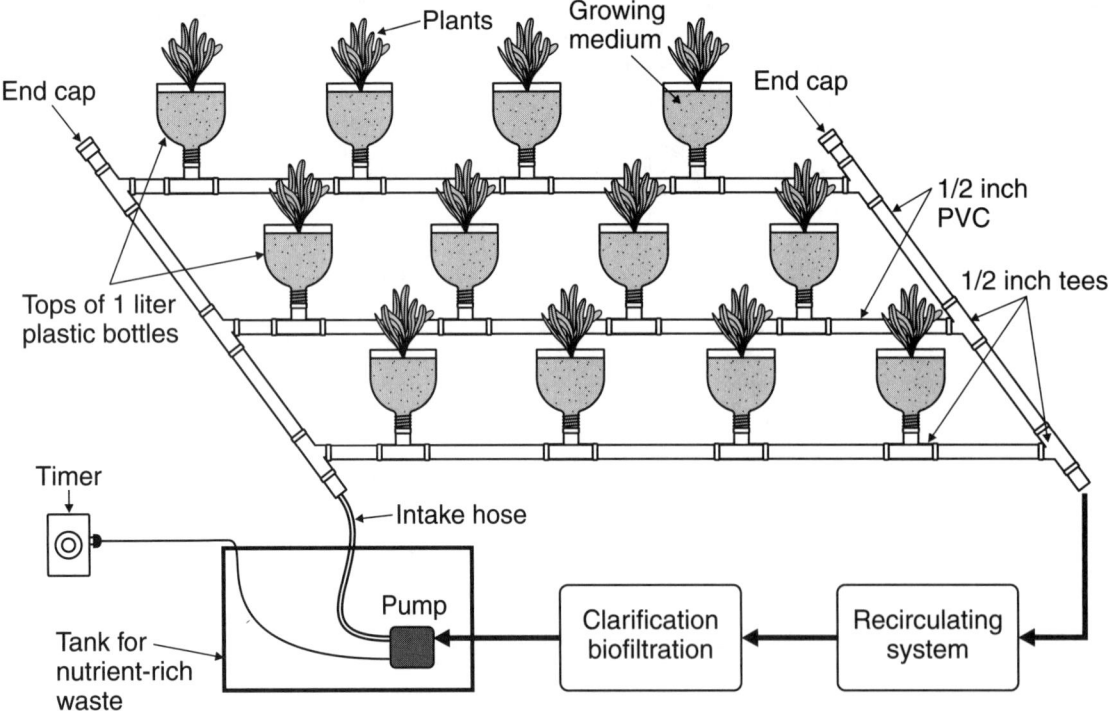

FIGURE 17-3 Example of a possible flow-through system that uses PVC pipe and the tops of 1-liter plastic bottles.

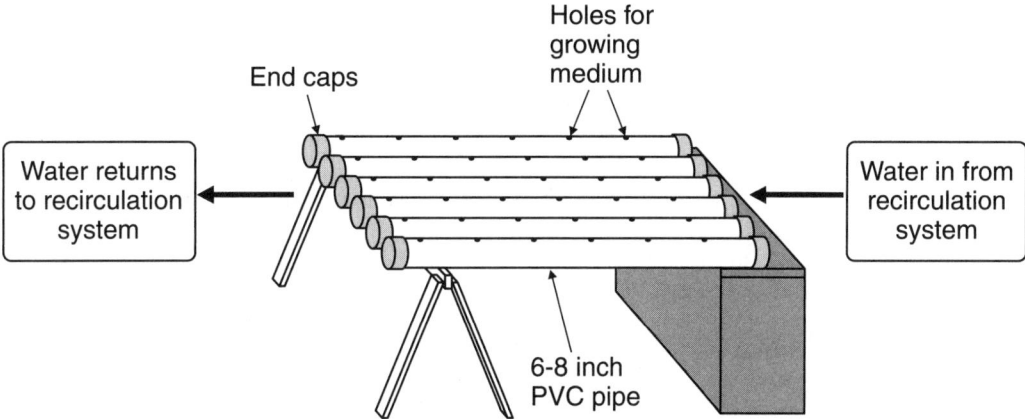

FIGURE 17-4 Example of a possible flow-through system constructed with large PVC pipe that could be combined with a recirculation system.

4. Place fiberglass screen over the top of the gravel. Fold the excess over or you can trim the screen with a pair of scissors. The screen acts as a filter to keep the growing medium in place. So try to fit the screen as close to the sides of the bucket as you can. If too much growing medium gets through the screen, it can actually clog the fill/drain tube.

5. Add the growing medium.

6. Fill the nutrient-rich bucket from the clarified, biofiltered, water from the fish tank and presoak the growing medium.

7. Plant seedling, rooted cutting, or seed in the growing medium.

8. Moving water through the system is accomplished by raising and lowering buckets or by the use of pumps (see Figure 17-5).

9. Maintain a daily log of the construction and operation of the system (Table 17-1).

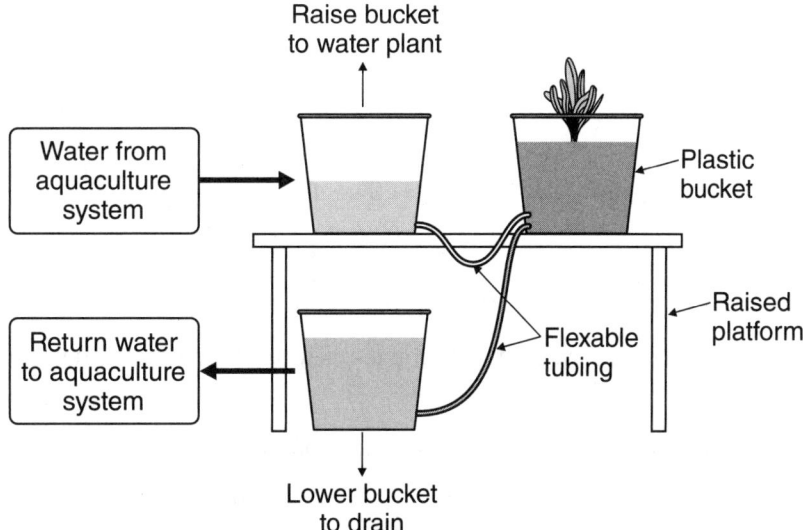

FIGURE 17-5 A simple gravity-flow system that could be combined with an aquarium or a recirculation system. The system could be modified to use pumps.

Care

1. Feed the fish twice daily and check for overfeeding.

2. Clean the tank and piping as necessary. Some algae is good, but too much clogs the pipes and can take over the tank.

3. Observe the plant. Some adjustments may be made in the nutrient content by adding fertilizer used in hydroponic systems.

4. Monitor the pH of the system.

5. Observe the health of the plants and the fish in the system.

RESOURCES

For more design information and operational information, the following websites and organization can be helpful.

Aquaponics
http://www.aquaponics.com

The Growing Edge
http://www.growingedge.com

Hydroponics Online
http://www.hydroponicsonline.com/freestuff.html

Esoteric Hydroponics
http://www.blunt.co.uk/manual/index.html

Hydroponic Society of America
P.O. Box 3075
San Ramon, CA 94538

ANALYSIS

Write a lab report based on the notes made in your daily log (Table 17-1). Respond to the remaining Analysis items in your lab report.

1. Examine the value of a nitrogen source to plant growth.

2. List the elements, besides nitrogen, necessary for plant growth.

3. Compare the pH for growing plants to that needed to successfully raise fish.

4. Plants need light for growth, so why should the aquarium or recirculation systems be shielded from light?

5. Discuss some advantages of hydroponics and of aquaponics.

6. Describe the role of a growing medium in a hydroponic or aquaponic system and give two examples.

Aquaculture Reports and Interpretation

INTRODUCTION

Part of any business requires reading and interpreting information. Aquaculture is no different. From government reports, to industry newsletters to extension publications, successful aquaculturalists analyze data and information to glean knowledge applicable to their operation. In this lab, you will read and analyze information about catfish production. This information is collected and distributed by the National Agricultural Statistics Service of the United States Department of Agriculture (http://www.nass.usda.gov/).

The purpose of this lab is to develop skills in reading and interpreting information contained in reports.

CORRELATION

This lab and/or Lab 19 can be used with Chapter 17 of *Aquaculture Science*, 3rd Edition.

BACKGROUND

The National Agricultural Statistics Service prepares and publishes reports from monthly surveys of catfish processed, end-of-the-month inventories, prices paid to catfish producers, prices received by processors and an annual catfish production report released the end of January each year.

MATERIALS

> ➤ Paper and pencil

> ➤ Calculator (optional)

(**Note:** A word processor and spreadsheet program may be substituted for the paper and pencil.)

PROCEDURES AND ANALYSIS

Study the following tables reproduced from the Catfish Production report released by the USDA National Agricultural Statistics Service (NASS) on January 2010.

1. Which state had the largest number of acres of ponds under construction in the year 2009?

2. Which state took the most acres of ponds out of production in the year 2009?

3. Which state increased the most in the number of broodfish in one year?

NASS
FACT FINDERS FOR AGRICULTURE
UNITED STATES DEPARTMENT OF AGRICULTURE

Washington, D.C.

Catfish Production

Released January 29, 2010, by the National Agricultural Statistics Service (NASS), Agricultural Statistics Board, U.S. Department of Agriculture. For information on *Catfish Production* call Chris Hawthorn at (202) 720-0585, office hours 7:30 a.m. to 4:00 p.m. ET.

Catfish Value of Sales Down 9 Percent from 2008

Catfish growers in the U.S. had sales of 373 million dollars during 2009, down 9 percent from 410 million dollars the previous year. The top four States (Mississippi, Alabama, Arkansas, and Texas) accounted for 93 percent of the U.S. total sales. The U.S. sales total of all foodsize fish decreased by 10 percent from 2008 to 352 million dollars in 2009. Fingerling and fry sales totaled 13 million dollars, an increase of 7 percent from 2008. Sales of stockers totaled 7 million dollars in 2009, compared to 8 million dollars in 2008.

By point of first sale, direct sales to processors accounted for 94 percent of total foodsize fish sales compared with 95 percent from the previous year. Direct sales to other producers accounted for 87 percent of stocker sales compared with 89 percent in 2008.

Water Surface Acres Down 22 Percent from January 1, 2009

The water surface acres being used for catfish production as of January 1, 2010 totaled 115 thousand acres, down 22 percent from the 147 thousand acres used a year earlier. Of the total acres, 3 thousand are to be renovated during the period of January 1 to June 30, 2010. An additional 140 acres are under construction or expected to be constructed and in use by July 1, 2010. During the period of July 1 through December 31, 2009, the area taken out of production totaled 10 thousand acres. As of January 1, 2010, foodsize fish were produced on 97 thousand acres, fingerling-producing acres totaled 14 thousand, and 2 thousand acres were being used for broodfish production.

Inventory Numbers

Catfish producers had 544 thousand **broodfish** on hand January 1, 2010, down 23 percent from January 1, 2009. **Large foodsize** fish on hand totaled 9 million on January 1, 2010, an 8 percent decrease from a year ago. The number of **medium foodsize** fish decreased 13 percent to 92 million, while **small foodsize** fish numbers decreased 13 percent to 169 million. **Large stockers** on hand January 1, 2010, at 152 million fish, were down 41 percent from the previous year. **Small stocker** numbers decreased 35 percent to 213 million. There were 430 million **fingerlings** on hand January 1, 2010, down 41 percent from January 1, 2009.

Catfish: Number of Operations and Water Surface Acres Used for Production, 2009-2010, and Total Sales, 2008-2009, by State and United States

State	Number of Operations on Jan 1		Water Surface Acres Used for Production During Jan 1 - Jun 30		Total Sales	
	2009 [1]	2010 [1]	2009	2010	2008	2009
	Number	*Number*	*Acres*	*Acres*	*1,000 Dollars*	*1,000 Dollars*
AL			22,100	19,800	93,254	90,688
AR			25,000	19,200	64,263	44,914
CA			2,400	1,500	7,913	8,074
LA			6,300	1,800	11,883	8,395
MS			80,200	64,000	206,288	196,787
NC			2,200	1,900	7,221	5,495
TX			3,800	2,900	13,212	12,644
Oth Sts [2]			4,900	3,700	5,964	5,570
US	1,306	994	146,900	114,800	409,998	372,567

[1] State level number of operations will only be published every five years in conjunction with the Census of Agriculture.

[2] Other States include State estimates not shown and States suppressed due to disclosure.

Catfish: Water Surface Acre Usage, by State and United States, 2008-2010

State	Acres Intended for Utilization During January 1- June 30										Acres Taken Out of Production During July 1- December 31	
	Foodsize		Fingerlings		Broodfish		Currently Under or Scheduled for:					
							Renovation		New Construction			
	2009	2010	2009	2010	2009	2010	2009	2010	2009	2010	2008	2009
	Acres	*Acres*	*Acres*	*Acres*	*Acres*	*Acres*	*Acres*	*Acres*	*Acres*	*Acres*	*Acres*	*Acres*
AL	21,100	19,200	480	380	320	120	180	120	85	30	1,000	370
AR	21,300	16,600	2,800	2,200	250	250	700	570			2,500	2,200
CA	1,700	1,100	290	190	100	80	140	70	320	*	*	*
LA	4,900	1,700	480	50			640	10			1,300	2,800
MS	64,000	52,000	12,100	9,700	1,400	1,300	2,800	2,100		50	6,700	3,500
NC	2,000	1,600	160	200	70	50	150	90	*	*	55	40
TX	3,100	2,600	420	190	220	70	205	85	75	50	550	135
Oth Sts [1]	2,500	1,900	1,700	1,300	340	370	200	65	40	10	660	840
US	120,600	96,700	18,430	14,210	2,700	2,240	5,015	3,110	520	140	12,765	9,885

* Not published to avoid disclosure of individual operations.

[1] Other States include State estimates not shown and States suppressed due to disclosure.

**Catfish: Inventory by Size Category, by State
and United States, January 1, 2009-2010**

Size Category and State	Number of Fish		Live Weight			
			Total		Average	
	2009	2010	2009	2010	2009	2010
	1,000	*1,000*	*1,000 Pounds*	*1,000 Pounds*	*Pounds per Fish*	*Pounds per Fish*
Large Foodsize						
AL	2,200	2,400	7,750	8,500	3.5	3.5
AR	1,290	1,190	4,550	4,300	3.5	3.6
CA	*	100	*	340	*	3.4
LA	430	380	1,490	1,300	3.5	3.4
MS	4,290	3,800	15,200	12,900	3.5	3.4
NC	260	230	825	740	3.2	3.2
TX	530	340	1,760	1,190	3.3	3.5
Oth Sts [1]	316	130	1,121	442	3.5	3.4
US	9,316	8,570	32,696	29,712	3.5	3.5
Medium Foodsize						
AL	23,500	30,400	50,000	56,400	2.1	1.9
AR	14,300	11,300	28,100	21,700	2.0	1.9
CA	*	270	*	540	*	2.0
LA	3,210	830	7,000	1,670	2.2	2.0
MS	57,100	42,700	111,300	82,800	1.9	1.9
NC	1,570	1,900	3,050	4,000	1.9	2.1
TX	4,140	3,440	8,050	6,700	1.9	1.9
Oth Sts [1]	1,790	750	3,380	1,440	1.9	1.9
US	105,610	91,590	210,880	175,250	2.0	1.9
Small Foodsize						
AL	37,400	42,500	39,300	44,800	1.1	1.1
AR	29,300	19,400	30,900	19,700	1.1	1.0
CA	890	1,980	943	1,790	1.1	0.9
LA	2,820	1,300	2,800	1,220	1.0	0.9
MS	114,000	94,000	119,500	93,500	1.0	1.0
NC	2,540	3,040	2,850	3,850	1.1	1.3
TX	5,170	5,190	4,500	5,050	0.9	1.0
Oth Sts [1]	1,750	1,210	1,630	1,010	0.9	0.8
US	193,870	168,620	202,423	170,920	1.0	1.0
Total Foodsize						
AL	63,100	75,300	97,050	109,700		
AR	44,890	31,890	63,550	45,700		
CA	1,624	2,350	2,593	2,670		
LA	6,460	2,510	11,290	4,190		
MS	175,390	140,500	246,000	189,200		
NC	4,370	5,170	6,725	8,590		
TX	9,840	8,970	14,310	12,940		
Oth Sts [1]	3,122	2,090	4,481	2,892		
US	308,796	268,780	445,999	375,882		

* Not published to avoid disclosure of individual operations.
[1] Other States include State estimates not shown and States suppressed due to disclosure.

**Catfish: Inventory by Size Category, by State
and United States, January 1, 2009-2010**

Size Category and State	Number of Fish		Live Weight			
			Total		Average	
	2009	2010	2009	2010	2009 [1]	2010
	1,000	*1,000*	*1,000 Pounds*	*1,000 Pounds*	*Pounds per 1,000 Fish*	*Pounds per 1,000 Fish*
Large Stockers						
AL	40,700	21,000	16,900	10,200	415	486
AR	27,500	19,100	11,400	7,550	415	395
CA	*	*	*	*	*	*
LA	2,820	*	1,120	*	397	*
MS	179,000	103,000	67,900	44,200	379	429
NC	4,870	5,070	2,800	3,150	575	621
TX	*	*	*	*	*	*
Oth Sts [2]	3,750	4,020	1,216	1,395	324	347
US	258,640	152,190	101,336	66,495	392	437
Small Stockers						
AL	32,600	25,700	3,200	2,900	98.2	112.8
AR	38,100	33,100	4,650	3,250	122.0	98.2
CA	*	890	*	120	*	134.8
LA	4,580	*	475	*	103.7	*
MS	234,000	144,000	24,600	14,100	105.1	97.9
NC	11,400	350	1,640	38	143.9	108.6
TX	*	*	*	*	*	*
Oth Sts [2]	6,749	9,140	670	808	99.3	88.4
US	327,429	213,180	35,235	21,216	107.6	99.5
Total Stockers						
AL	73,300	46,700	20,100	13,100		
AR	65,600	52,200	16,050	10,800		
CA	5,030	1,490	1,084	450		
LA	7,400	*	1,595	*		
MS	413,000	247,000	92,500	58,300		
NC	16,270	5,420	4,440	3,188		
TX	99	980	12	140		
Oth Sts [2]	5,370	11,580	790	1,733		
US	586,069	365,370	136,571	87,711		
Fingerlings						
AL	78,800	23,400	3,150	1,250	40.0	53.4
AR	74,700	44,200	3,000	1,700	40.2	38.5
CA	4,500	5,160	129	190	28.7	36.8
LA	7,710	*	289	*	37.5	*
MS	511,000	323,000	14,100	10,500	27.6	32.5
NC	5,530	8,000	139	230	25.1	28.8
TX	25,100	10,300	1,190	490	47.4	47.6
Oth Sts [2]	21,000	15,530	683	685	32.5	44.1
US	728,340	429,590	22,680	15,045	31.1	35.0

* Not published to avoid disclosure of individual operations.
[1] Revised.
[2] Other States include State estimates not shown and States suppressed due to disclosure.

**Catfish: Inventory by Size Category, by State
and United States, January 1, 2009-2010**

Size Category and State	Number of Fish		Live Weight			
			Total		Average	
	2009	2010	2009	2010	2009	2010
	1,000	*1,000*	*1,000 Pounds*	*1,000 Pounds*	*Pounds per Fish*	*Pounds per Fish*
Broodfish						
AL	39	10	197	73	5.1	7.3
AR	70	50	352	212	5.0	4.2
CA	19	17	89	90	4.7	5.3
LA		*		*		*
MS	420	350	2,150	1,900	5.1	5.4
NC	10	11	63	75	6.3	6.8
TX	86	20	424	120	4.9	6.0
Oth Sts [1]	60	86	265	478	4.4	5.6
US	704	544	3,540	2,948	5.0	5.4

* Not published to avoid disclosure of individual operations.

[1] Other States include State estimates not shown and States suppressed due to disclosure.

Catfish: Sales by Size Category, by State and United States, 2008-2009

Size Category and State	Number of Fish		Live Weight Total		Live Weight Average		Sales Total		Sales Average price per pound	
	2008	2009	2008	2009	2008	2009	2008	2009	2008	2009
	1,000	*1,000*	*1,000 Pounds*	*1,000 Pounds*	*Pounds per fish*	*Pounds per fish*	*1,000 Dollars*	*1,000 Dollars*	*Dollars*	*Dollars*
Foodsize										
AL	82,600	66,600	131,600	128,900	1.6	1.9	92,120	90,230	0.70	0.70
AR	51,100	31,200	83,700	58,100	1.6	1.9	62,775	42,994	0.75	0.74
CA	2,220	2,550	3,150	3,400	1.4	1.3	7,592	7,820	2.41	2.30
LA	7,420	4,660	15,400	11,500	2.1	2.5	11,827	8,395	0.77	0.73
MS	143,000	146,000	252,370	249,000	1.8	1.7	191,801	181,770	0.76	0.73
NC	4,040	3,120	8,050	6,150	2.0	2.0	6,843	5,166	0.85	0.84
TX	11,100	10,500	16,900	16,100	1.5	1.5	12,844	12,558	0.76	0.78
Oth Sts [1]	2,530	1,680	3,750	2,800	1.5	1.7	3,488	3,080	0.93	1.10
US	304,010	266,310	514,920	475,950	1.7	1.8	389,290	352,013	0.76	0.74
	1,000	*1,000*	*1,000 Pounds*	*1,000 Pounds*	*Pounds per 1,000 fish*	*Pounds per 1,000 fish*	*1,000 Dollars*	*1,000 Dollars*	*Dollars*	*Dollars*
Stockers										
AL	3,700	500	262	280	71	560	472	202	1.80	0.72
AR	*	*	*	*	*	*	*	*	*	*
CA	135		76		563		219		2.88	
LA										
MS	60,200	40,000	6,000	4,500	100	113	6,060	4,635	1.01	1.03
NC	*	*	*	*	*	*	*	*	*	*
TX	*	130	*	20	*	154	*	22	*	1.08
Oth Sts [1]	10,440	17,790	1,574	2,657	151	149	1,587	2,626	1.01	0.99
US	74,475	58,420	7,912	7,457	106	128	8,338	7,485	1.05	1.00
Fingerlings & Fry										
AL	11,100	3,800	210	95	18.9	25.0	582	253	2.77	2.66
AR	52,300	29,100	453	272	8.7	9.3	761	541	1.68	1.99
CA	370	1,300	16	73	43.2	56.2	65	222	4.08	3.04
LA	1,540		31		20.1		56		1.80	
MS	207,000	202,000	5,800	7,100	28.0	35.1	8,352	10,295	1.44	1.45
NC	3,540	3,200	200	190	56.5	59.4	356	329	1.78	1.73
TX	3,170	1,270	167	65	52.7	51.2	212	62	1.27	0.96
Oth Sts [1]	20,900	14,700	544	481	26.0	32.7	1,692	1,193	3.11	2.48
US	299,920	255,370	7,421	8,276	24.7	32.4	12,076	12,895	1.63	1.56

* Not published to avoid disclosure of individual operations.
[1] Other States include State estimates not shown and States suppressed due to disclosure.

**Catfish: Sales by Size Category, by State
and United States, 2008-2009**

Size Category and State	Number of Fish		Live Weight				Sales			
			Total		Average		Total		Average price per pound	
	2008	2009	2008	2009	2008	2009	2008	2009	2008	2009
	1,000	*1,000*	*1,000 Pounds*	*1,000 Pounds*	*Pounds per Fish*	*Pounds per Fish*	*1,000 Dollars*	*1,000 Dollars*	*Dollars*	*Dollars*
Broodfish										
AL	7	*	36	*	5.1	*	80	*	2.21	*
AR	*	*	*	*	*	*	*	*	*	*
CA	*	2	*	13	*	6.5	*	32	*	2.46
LA		*		*		*		*		*
MS	39	25	112	100	2.9	4.0	75	87	0.67	0.87
NC	*	*	*	*	*	*	*	*	*	*
TX	*	1	*	2	*	2.0	*	2	*	1.17
Oth Sts [1]	29	15	138	69	4.8	4.6	139	53	1.01	0.77
US	75	43	286	184	3.8	4.3	294	174	1.03	0.95

* Not published to avoid disclosure of individual operations.

[1] Other States include State estimates not shown and States suppressed due to disclosure.

Catfish: Foodsize and Stockers Percent Sold by Point of First Sale,
by State and United States, 2008-2009

Size Category and State	Processor		Fee-Fishing and Recreational Use		Live Haulers		Retail		Other	
	2008 [1]	2009	2008 [1]	2009	2008	2009	2008 [1]	2009	2008	2009
	Percent	*Percent*	*Percent*	*Percent*	*Percent*	*Percent*	*Percent*	*Percent*	*Percent*	*Percent*
Foodsize										
AL	97	95	*	1	1	1	*	1	2	2
AR	96	95	*	*	3	5	*	*	*	*
CA			10	13	11	22	73	53	6	12
LA	98	95	1	5			*			
MS	100	99	*		*		*		*	1
NC	96	95	*	*	*	*	*	*	1	3
TX	83	83	1	1	10	15			6	1
Oth Sts [2]	20	21	33	43	30	24	12	4	6	8
US [3]	95.3	93.9	0.6	1.1	1.6	2.1	1.6	1.5	0.9	1.4

	Other Producers		Fee-Fishing and Recreational Use		Live Haulers		Government		Other	
	2008	2009	2008	2009	2008	2009	2008	2009	2008	2009
	Percent	*Percent*	*Percent*	*Percent*	*Percent*	*Percent*	*Percent*	*Percent*	*Percent*	*Percent*
Stockers										
AL	96	97	1	1	2	1			1	1
AR	62	89			*	11	*		*	
CA	*		*		7		*		*	
LA										
MS	100	99	*						*	1
NC	*	*	*	*						
TX	95	88	5	*		*				
Oth Sts [2]	29	40	4	1	58	57	4	1	4	1
US [3]	89.0	86.9	0.8	0.2	8.9	11.9	0.6	0.2	0.7	0.8

* Not published to avoid disclosure of individual operations or data are less than 0.5 percent.
[1] Revised.
[2] Other States include State estimates not shown and States suppressed due to disclosure.
[3] Sum of US Point of First Sale by outlet may not add to 100 percent.

Terms and Definitions Used for Catfish Production Estimates

Broodfish - Fish kept for egg production, including males. Broodfish produce the fertilized eggs, which go to hatcheries. The most desirable size is 3 to 10 pounds or 4 to 6 years of age.

Fingerlings - Fish weighing 2 to 60 pounds per 1,000 fish or 2 to 6 inches in length.

Foodsize (large) - Fish weighing over 3 pounds.

Foodsize (medium) - Fish weighing over one and one-half pounds to 3 pounds.

Foodsize (small) - Fish weighing over three-fourths of a pound to one and one-half pounds.

Fry - Fish weighing less than 2 pounds per 1,000 fish or less than 2 inches in length.

Stockers (large) - Fish weighing over 180 pounds to 750 pounds per 1,000 fish.

Stockers (small) - Fish weighing over 60 pounds to 180 pounds per 1,000 fish or over 6 inches in length.

4. What is the average weight ranges for the following:

   Small foodsize   _____

   Broodfish   _____

   Small stockers   _____

5. Between the two years in the report, name two states that dropped in the live weight of large, food-size fish.

6. In the most recent year, how many large food-size and medium food-size fish did the following states have at the time of the report?

   Alabama   _____   _____

   Mississippi   _____   _____

   Louisiana   _____   _____

   Texas   _____   _____

7. Which state sold the most foodsize catfish for fee-fishing and recreational use?

8. Name the state that lists the largest number of broodfish.

9. Determine what percentage of all food size catfish in the United States were in the following states in 2009:

   Alabama   _____

   California   _____

   Mississippi   _____

   Texas   _____

10. Which state produces the most fingerlings and what percentage of the total fingerlings in the United States does this state produce?

11. How many live weight tons of large foodsize catfish were produced in the United States in 2010?

12. Which five states produced the most of these large foodsize catfish in 2010?

13. In 2009, how much would a producer have received for 5,000 pounds of medium food-size catfish?

14. What was the average, foodsize, live weight price per pound paid to catfish producers in 2009, and how much did these fish weigh, on the average?

15. According to the data for the most recent year, what percentage of all the surface acres of water in the United States does Mississippi represent?

16. In 2010, what was the total value of catfish sales (all sizes) in the United States?

INTRODUCTION

Fish should be sample counted at least monthly to ensure that they are growing as expected and to keep track of loading rates. Feeding according to a feeding rate chart allows a check of daily ration amounts and adjustments as necessary. A sample of fish is netted into a bucket of water suspended from a spring tension scale. The weight is recorded and the number of fish is determined as they are poured back into the tank. If fish are graded rather uniformly, three or four samples from different areas are sufficient. Fish size (expressed as number per pound) is calculated by dividing the number of fish in each sample by the total sample weight. The average for each tank is then used to estimate the weight of fish in the entire raceway.

The purpose of this lab is to teach the method used to estimate the number of fish in a group and to estimate a certain number of fish from a large group.

CORRELATION

This lab and/or Lab 18 can be used with Chapter 17 of *Aquaculture Science*, 3rd Edition.

BACKGROUND

Fish producers often need to estimate numbers of fish. Estimates are used because fish, especially fingerlings (1- to 2-in. fish), are difficult to count. For example, a fish producer may have a tank of fingerling catfish to sell and needs to know how many fish it contains. Fingerlings are frequently sold by the fish instead of by weight. Likewise, if a fish producer wants to stock a certain number of fish into a pond, he or she may need to measure out the number needed from a large batch of fish. Because the desired number of fish is often in the thousands or hundreds of thousands, it is impossible to count all the fish. To estimate large numbers of fish, producers count a small group of fish and weigh them. Producers will then know how much a certain number of fish weigh and can use this knowledge to weigh out larger numbers of fish.

An Example

A fish producer wants to estimate how many fingerling catfish are in his hatchery tank. Then he wants to stock a 2-acre pond at the rate of 1,000 fingerlings per acre. First he counts out a sample of 100 fish from the hatchery tank, weighs them, and finds that the 100 fish together weigh 0.5 lb. Next, he weighs all the fish in the vat. He does this by filling a tub half full with water and weighing the tub and water. Then he uses a dip net to remove all the fish from the tank, places the fish into the half-filled tub and weighs the tub again. By subtracting the first weight (the weight without fish) from the second weight (the weight with fish), he knows how much all the fish weigh. After doing this, he discovers that he has 30 lbs of fish in the tub. The relationship between the sample measurements and the fish in the tub can be described as follows:

$$\frac{\text{Sample Number}}{\text{Sample Weight}} = \frac{\text{Total Number}}{\text{Total Weight}}$$

Using this relationship, the fish producer cross-multiplies and estimates that he has 6,000 fish in the tub.

$$\frac{100 \text{ fish}}{0.5 \text{ lb.}} = \frac{? \text{ fish}}{30 \text{ lbs.}} \rightarrow \frac{100 \times 30}{0.5} = 6,000 \text{ fish}$$

Next, the producer wants to measure out enough fish to stock a 2-acre pond at the rate of 1,000 fish per acre. He multiplies his acreage by the desired number of fish per acre and determines that he needs to measure out 2,000 fish.

The producer does not want to count 2,000 individual fish, so he will measure out approximately 2,000 fish by weighing them. Using the following formula, he can find out how much 2,000 fish will weigh.

He uses the same formula as before but substitutes 2,000 into the total numbers portion of the equation and cross-multiplies to calculate the weight of 2,000 fish.

$$\frac{100 \text{ fish}}{0.5 \text{ lb.}} = \frac{2000 \text{ fish}}{? \text{ lbs.}} \rightarrow \frac{2000 \times 0.5}{100} = 10 \text{ lbs}$$

MATERIALS

➤ One quart jar of navy beans

➤ 8-in. × 8-in. pan

➤ 8-in. × 13-in. pan

➤ Scales

➤ Pencil

➤ Calculator (optional)

PROCEDURES

1. Beans will be used to illustrate how to estimate large numbers of fish or other items that are not easily counted. In this exercise, you will estimate the number of beans in a quart jar in the same manner that the fish producer estimated the number of fish in the tub. (See Figure 19-1.)

2. Weigh an empty jar on the scale and record the weight in Table 19-1 of the Analysis section. Then, fill the jar with beans and weigh the jar and beans. Record your answer in Table 19-1. Subtract the weight of the empty jar from the weight of the full jar to get the weight of the beans. Record your answer in Table 19-1.

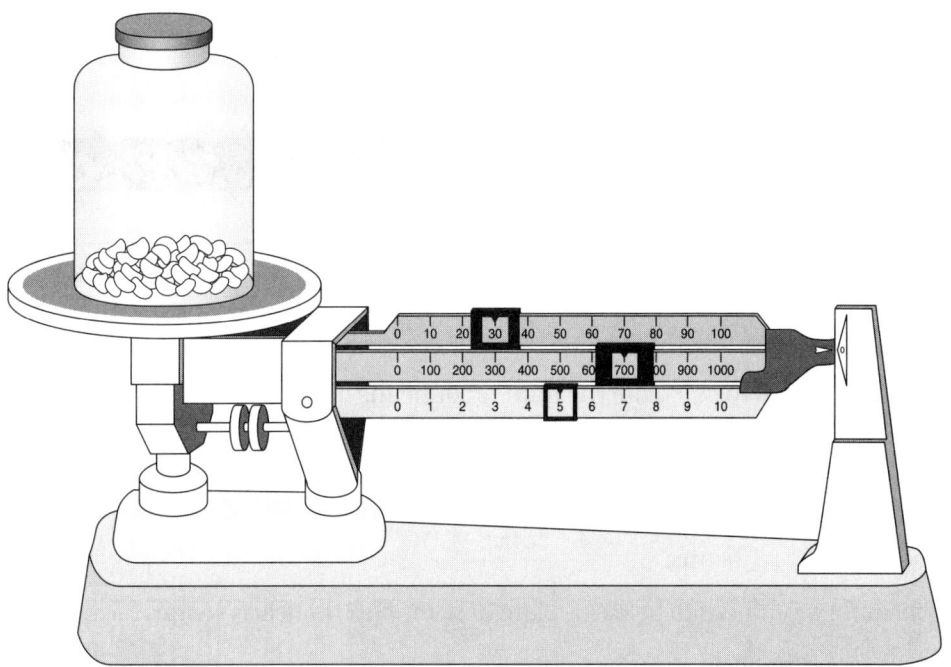

FIGURE 19-1 Weighing a sample of beans is the first step in calculating the number of beans in a jar.

3. Count out 50 beans from the jar and weigh them. Now you know:
   - The weight of all the beans
   - The weight of the sample of beans
   - The number of beans in the sample

   With this information, you can plug in the missing figures in the formula and solve it mathematically to estimate the number of beans in the jar. Complete item 2 in the Analysis section.

4. In this part of the exercise, the beans will represent fingerlings used to stock ponds. The baking pans will represent ponds. Imagine that an 8-in. × 8-in. baking pan represents a 1-acre pond. Since an 8-in. × 8-in. pan contains 64-in^2, you can assume that every 64 in^2 equals 1 acre. An 8-in. × 13-in. pan contains 117 in^2. Since 117 divided by 64 in^2 equals 1.8, you can say this pan represents a 1.8-acre pond. In this exercise, you will stock the two pans at the rate of 300 beans (fish) per acre.

5. First, you will stock the 8-in. × 8-in. pan. Because this pan represents exactly one acre, you will need to weigh out 300 beans.

6. Now compare the two pans. Do the beans look like they are stocked at the same density in both pans? They should. If they do not look like the same densities, you may have made a math error. Double check the math and try again. Complete Table 19-2 in the Analysis section.

7. Count the beans in the two pans to see how close you came to the number you wanted. Record the results in Table 19-2.

ANALYSIS

1. Complete Table 19-1 to determine the weight of 1 qt of beans.

TABLE 19-1 DETERMINING THE WEIGHT OF ONE QUART OF BEANS	
Weight of the jar with beans	
Weight of the empty jar	
Weight of the beans	

2. Complete this formula to calculate the number of beans in the jar:

$$\frac{\text{Sample Number}}{\text{Sample Weight}} = \frac{\text{Total Number}}{\text{Total Weight}} \rightarrow \underline{\quad} = \frac{?}{\underline{\quad}} \rightarrow \frac{\times}{\underline{\quad}} \rightarrow \text{beans}$$

The jar contains _____ beans.

3. Use the formula given here to help calculate how much 300 beans weigh. Record the result in Table 19-2.

$$\frac{\text{Sample Number}}{\text{Sample Weight}} = \frac{\text{Total Number}}{\text{Total Weight}} \rightarrow \underline{\quad} = \frac{\underline{\quad}}{?} \rightarrow \frac{\times}{\underline{\quad}} \rightarrow \text{(weight)}$$

Slowly add beans to the scale until it reads the weight calculated in the formula. Now spread these beans evenly in the bottom of the 8-in. × 8-in. pan.

4. Use the formula given here to calculate how much 540 beans weigh. Record the result in Table 19-2.

$$\frac{\text{Sample Number}}{\text{Sample Weight}} = \frac{\text{Total Number}}{\text{Total Weight}} \rightarrow \underline{\quad} = \frac{\underline{\quad}}{?} \rightarrow \frac{\times}{\underline{\quad}} \rightarrow \text{(weight)}$$

Slowly add beans to the scale until it reads the weight calculated in the formula. Now spread these beans evenly in the bottom of the 8-in. × 13-in. pan.

5. Record the results from the pond stocking exercise in Table 19-2.

6. Do the pans look like they have the same density of beans?

7. Based on your results in this exercise, do you think that estimating numbers by weight is a reliable method? Explain.

8. Name other situations in which this method of estimating would be useful.

TABLE 19-2 STOCKING RATE OF FINGERLINGS IN PONDS				
Pan Size	Represents Pond Size	Recommended Stocking Rate for This Pond	Weight Needed	Actual Count
8-in. × 8-in. pan	1 acre	300		
8-in. × 8-in. pan	1.8 acres	540		

LAB 20 — Using the Internet to Find a Career or Job

INTRODUCTION

One purpose of education and learning is to become employable and stay employable—to get and keep a job. People look for careers, and careers look for people. Two broad categories of career opportunities are working for someone else and working for yourself. The industry of agriculture, food, and natural resources includes a wide variety of career opportunities. These opportunities need to be found and matched with the motivations, abilities, skills, and knowledge of individuals. This process is called career exploration (see Figure 20-1).

The purpose of this lab is to learn how to find potential careers/jobs and match these careers/jobs with the individual.

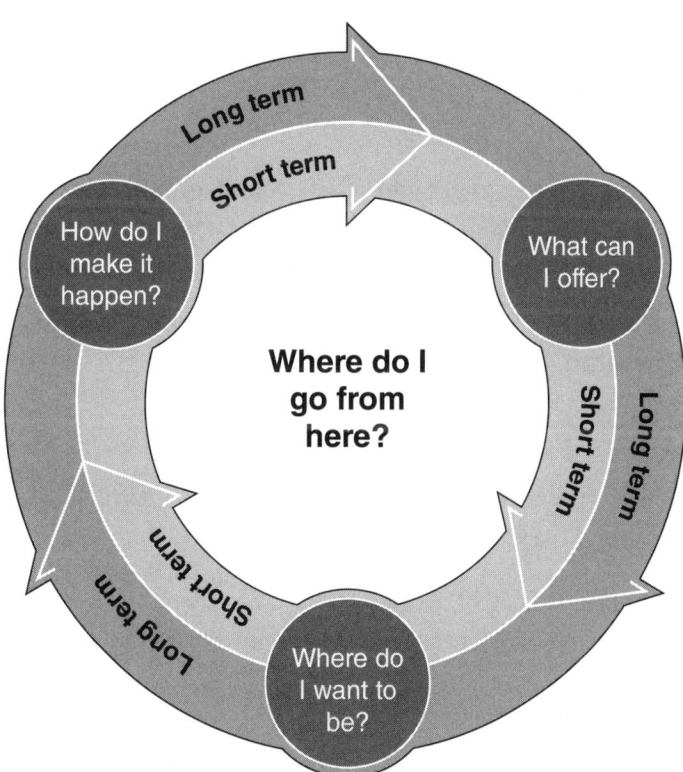

FIGURE 20-1 The cycle and steps of career exploration.

CORRELATION

This lab can be used with Chapter 18 of *Aquaculture Science*, 3rd Edition.

BACKGROUND

Some companies and services maintain computerized databases of jobs. Interested individuals use the Internet to search for jobs that match their qualifications and desires. Often, resumés can be submitted over the Internet. This type of job listing opens the door wide to potential jobs, but often not local jobs.

Increasingly, career development centers are using the Internet as a resource for career planning. Major career planning activities include self-assessment and career exploration. The goal of career exploration is to help you to find opportunities that match your personal and professional needs.

MATERIALS

➤ Internet access

➤ Word processor

➤ Printer

➤ Individual e-mail address

PROCEDURES

1. Use the following website to complete a personal assessment title "Motivational Appraisal of Personal Potential" (individual email address needed to receive results): www.assessment.com/MAPPMembers/TakeMAPP.asp (see Figure 20-2).

2. Visit the following websites to explore job/career opportunities that are posted online and cut and paste six jobs of interest (at least three from different websites) into a word processing document:

   • Aquaculture Jobs: www.aquaculturejobs.com

   • AgriSeek: www.agriseek.com

   • Agricultural Classified Ads: www.agriads.com

   • AgCareers.com: www.agcareers.com

   • Produce Careers: www.producecareers.com

   • Produce Careers (large companies): www.blueskysearch.com

   • America's Job Bank: http://www.jobbankinfo.org/

   • Monster: jobsearch.monster.com

3. Using the newsletter archives on the AgCareers.com website (www.agcareers.com) find one tip or piece of advice that is useful to you as a jobseeker. Articles in the archives include: salary negotiation, writing resumes and cover letters, and the overall job search process.

4. Choose one employer from your search in #2 above. Assume that you have an interview with this employer—before the interview you will need to learn as much as possible about that prospective employer. Use the Internet as a tool to assist you in finding up-to-date information about the employer/company. Internet Business Directories such as the World Wide Yellow Pages (www.yellow.com) are resources for helping you to find information on a company. Often the company website or a phone call will answer the question.

FIGURE 20-2 For the individual with the right motivation, scientific research in aquaculture makes a great career choice. (Source: ARS)

ANALYSIS

1. Write a one-page analysis of the report from the results of your "Motivational Appraisal of Personal Potential (MAPP)," which will include:
   - Top ten jobs with your greatest potential for success
   - Interest in job content
   - Motivation to work with people
   - Preference to work with things

2. Turn in your one-page analysis attached to the report from the results of your "Motivational Appraisal of Personal Potential."

3. Write a one-page analysis of your search to find six job/careers on the websites listed in #2 under "Methods." Briefly describe why each of the six jobs/careers interested you. Include this one-page analysis with a printout of the six jobs/careers that you cut and pasted from the websites.

4. Report to the class on the useful tip or advice that you found on the AgCareers.com newsletter archive.

5. For the company/employer selected in #4 under "Methods," find answers to the following questions and turn in your report:

- What are the company's products and services?

- Who are the company's customers?

- What is the size of the company? Has the company grown over the last five years?

- Is the company profitable?

- Who are the companies major competitors?

- What are the work schedules? How many hours a day do employees work?

OTHER RESOURCES

NextStepU: www.nextstepu.com/

Career Builder: www.careerbuilder.com/

Occupational Outlook Handbook: www.bls.gov/OCO/

Job Web: www.jobweb.org

Monster College: college.monster.com

APPENDIX A Supply Companies

Carolina Biological Supply Company, PO Box 6010, Burlington, NC 27216-6010
http://www.carolina.com

Connecticut Valley Biological, 82 Valley Road, PO Box 326, Southampton, MA 01073
www.ctvalleybio.com

Environmental Test System, Inc., PO Box 4659, Elkhart, IN 46514
http://www.aquachek.com

Fisher Scientific, 485 Frontage Road, Burr Ridge, IL 60521
http://www.fishersci.com or http://www.fisheredu.com

HACH Company, PO Box 389, Loveland, CO 80539
http://www.hach.com

LAB-AIDS, Inc., 17 Colt Court, Ronkonkoma, NY 11779
http://www.lab-aids.com

LaMotte Company, PO Box 329, Chestertown, MD 21620
http://lamotte.com

NASCO, 901 Janesville Avenue, PO Box 901, Fort Atkinson, WI 53538-0901
www.enasco.com

Nebraska Scientific, 3823 Leavenworth Street, Omaha, NE 68105
http://www.nebraskascientific.com

Three Rivers of Brooksville, Inc., P.O. Box 10369, Brooksville, FL 34603, (352) 544-0333,
Fax (353) 848-0100
www.growafrog.com

Ward's Inc., 5100 West Henrietta Road, PO Box 92912, Rochester, NY 14692-9012
http://www.wardsci.com

APPENDIX B *Laboratory Rules for Students*

1. Do not mix chemicals or perform unscheduled experiments without the instructor's approval. Activities in the lab manual are safe if directions are followed.

2. Never use chemicals from an unlabeled container.

3. Do not taste chemicals or bring them into contact with eyes, nose, or mouth.

4. When live animals are used, demonstrate concern for their welfare.

5. Clean up spilled liquids immediately.

6. Discard any wastes in the appropriate container.

7. Keep your work area clean.

8. Treat other students with respect and avoid exposing them to danger.

9. Always read and follow directions. Never guess.

10. In case of an accident, notify your instructor immediately.

11. Keep a lab manual and make notes on all observations.

12. Read and follow all material safety data sheets that companies send with chemicals.

Hydrochloric Acid (HCl)

CAUTION: Do not inhale the vapors. Immediately clean up any spills. Generous amounts of water are needed to remove acid from skin and clothing. Dispense in dropper bottles. Never pour water into acid when diluting a concentrated solution!

Iodine (Tincture of Iodine). Use methyl alcohol. Dissolve 2 gm of iodine in 120 ml of alcohol. Dispense in dropper bottles.

CAUTION: Iodine crystals are irritating to the skin. Handle with care and label carefully.

Nitric Acid (HNO₃)

WARNING: Concentrated nitric acid is very corrosive to skin, clothing, and equipment. Use only a dropper bottle when dispensing the acid.

Phenol Red. Use distilled water. Dissolve 1 gm of sodium hydroxide pellets (NaOH) in 100 ml of water. Dissolve 1 gm of phenol red in 200 ml of water. These are the stock solutions from which you prepare the indicator solutions as needed. To use, add 4 drops of the NaOH solution to 20 ml of phenol red stock solution. Add enough water to make 2,000 ml of the phenol red indicator solution. Label all three containers. The phenol red can be made more sensitive by additional dilution.

Phenolphthalein (PHTH). (0.5%). Dissolve 1 gm of Phenolphthalein powder in 200 ml of 95 percent methyl alcohol.

Sodium Hydroxide (4.0%). Use distilled water. Dissolve 8.0 gm of sodium hydroxide pellets (NaOH) in 150 ml of water. Add water to make 200 ml of solution. Label the container.

Sodium Thiosulfate (Na₂S₂O₃). Dissolve 0.62 gm of solid $Na_2S_2O_3$ in 200 ml of distilled water. Dispense in dropper bottles.

Starch Solution. Put 4 gm of soluble potato starch into 200 ml of hot tap water. (Make a fresh solution each week.) Heat the solution to about 90°C. Stir thoroughly during heating.

Sulfuric Acid (H₂SO₄)

WARNING: Concentrated sulfuric acid is very corrosive. Handle with great care. Dispense with droppers.

Winkler's Solution A. Use distilled water. Dissolve 48 grams manganous sulfate ($MnSO_4$-$2H_2O$) in 80 ml of water. Filter. Add water to make 100 ml of solution. Dispense in dropper bottles. Label the container.

Winkler's Solution B. Use distilled water. Dissolve 140 grams of potassium hydroxide pellets (KOHOH) in 160 ml of water. Cool. Dissolve 30 grams of potassium iodide (KI) in the KOHOH solution. Add water to make 200 ml of solution. Label the container.

(Note: Prepare the solution in a plastic or heat-resistant glass container. Be very careful with this highly caustic solution. Use care in preparation and in use.)

To Convert English	To Metric, Multiply by	To Convert Metric	Multiply by	To Get English
acres	0.4047	hectares	2.47	acres
acres	4047	m^2	0.000247	acres
BTU	1055	joules	0.000948	BTU
BTU	0.0002928	kwh	3415.301	BTU
BTU/hr.	0.2931	watts	3.411805	BTU/hr.
bu.	0.03524	m^3	28.37684	bu.
bu.	35.24	l	0.028377	bu.
ft.^{3}	0.02832	m^3	35.31073	ft.^{3}
ft.^{3}	28.32	l	0.035311	ft.^{3}
in.^{3}	16.39	cm^3	0.061013	in.^{3}
in.^{3}	1.639×10^{-5}	m^3	61012.81	in.^{3}
in.^{3}	0.01639	l	61.01281	in.^{3}
yd.^{3}	0.7646	m^3	1.307873	yd.^{3}
yd.^{3}	764.6	l	0.001308	yd.^{3}
ft.	30.48	cm	0.032808	ft.
ft.	0.3048	m	3.28084	ft.
ft./min.	0.508	cm/sec.	1.968504	ft./min.
ft./sec.	30.48	cm/sec.	0.032808	ft./sec.
gal.	3785	cm	30.000264	gal.

(continued)

To Convert English	To Metric, Multiply by	To Convert Metric	Multiply by	To Get English
gal.	0.003785	m^3	264.2008	gal.
gal.	3.785	l	0.264201	gal.
gal./min.	0.06308	l/sec.	15.85289	gal./min.
in.	2.54	cm	0.393701	in.
in.	0.0254	m	39.37008	in.
mi.	1.609	km	0.621504	mi.
mph	26.82	m/min.	0.037286	mph
oz.	28.349	gm	0.035275	oz.
fl. oz.	0.02947	l	33.93281	fl. oz.
liq. pt.	0.4732	l	2.113271	liq. pt.
lb.	453.59	gm	0.002205	lb.
qt.	0.9463	l	1.056747	qt.
$ft.^2$	0.0929	m^2	10.76426	$ft.^2$
$yd.^2$	0.8361	m^2	1.196029	$yd.^2$
tons	0.9078	tonnes	1.101564	tons
yd.	0.0009144	km	1093.613	yd.
yd.	0.9144	m	1.093613	yd.

APPENDIX E

Fahrenheit to Centigrade Temperature Conversions[1]

°F	°C	°F	°C	°F	°C
100	37.8	81	27.2	62	16.7
99	37.2	80	26.7	61	16.1
98	36.7	79	26.1	60	15.6
97	36.1	78	25.6	59	15.0
96	35.6	77	25.0	58	14.4
95	35.0	76	24.4	57	13.9
94	34.4	75	23.9	56	13.3
93	33.9	74	23.3	55	12.8
92	33.3	73	22.8	54	12.2
91	32.8	72	22.2	53	11.7
90	32.2	71	21.7	52	11.1
89	31.7	70	21.1	51	10.6
88	31.1	69	20.6	50	10.0
87	30.6	68	20.0	49	9.4
86	30.0	67	19.4	48	8.9
85	29.4	66	18.9	47	8.3
84	28.9	65	18.3	46	7.8
83	28.3	64	17.8	45	7.2
82	27.8	63	17.2	44	6.7
43	6.1	39	3.9	35	1.7
42	5.6	38	3.3	34	1.1
41	5.0	37	2.8	33	0.6
40	4.4	36	2.2	32	0.0

[1]Formulas used: $°C = (°F - 32) \times 5/9$ or $°F = (°C \times 9/5) + 32$

APPENDIX F

Do's and Don'ts of Recirculating Systems

➤ Do monitor your water quality daily with several oxygen tests, both before and ½ to 1 hour after feeding.

➤ Do keep records of feed and growth of fish.

➤ Do keep records of water quality to see trends.

➤ Do keep a small quantity of floating feed to see if your fish are active and feeding.

➤ Do keep records of mortalities and fish removed from the system.

➤ Do make increases in feed amounts gradually.

➤ Do change feed sizes as your fish grow.

➤ Do allow makeup water that is chlorinated to stand for several days with an air stone in it before using in the system.

➤ Do be alert to changes in appetite or general behavior of the fish.

➤ Do have makeup water standing by for emergencies.

➤ Do handle fish gently and as little as possible.

➤ Do use some type of netting or screen over the tanks to prevent fish from jumping out of the tanks.

➤ Do make feed charts for others to follow.

➤ Do remember that decreases in feed will upset the biofilter as much as increases.

➤ Do remember that this is a learning process and failures sometimes result in more gain than successes.

➤ Do shield your tanks and biofilter from excessive light because of algae and light will inhibit bacterial metabolism.

➤ Do have a plan of action for power failures.

➤ Do generate as much help and interest from others as possible.

➤ Don't feed any moldy feed.

➤ Don't get excited if a fish or two die.

➤ Don't get in a hurry to stock your fish before the biofilter is active.

➤ Don't make rapid changes in any water quality parameter except oxygen.

➤ Don't expect clear water in your system; good quality water in a recirculation system will eventually look like good onion soup, without cheese and croutons.

➤ Don't get upset if your fish don't grow perfectly evenly.

APPENDIX G

Embryonic Development of Fish Eggs Using the Japanese Medaka

INTRODUCTION

Appendix G was originally one of the lab activities. Because this project may not fit for all schools it was moved to this appendix to be used as an additional or optional lab activity to Chapter 4 or 6 of *Aquaculture Science*, 3rd Edition.

Sexual reproduction is the process of creating new organisms of the same species through the union of the male and female sex cells—sperm and eggs. Males and females exist in most species. Testes in the males produce sperm. Ovaries in the females produce eggs or ova. Fertilization occurs when the sperm unites with the egg, forming a zygote. After a period of incubation, the zygote develops into a new organism. An understanding of the reproductive process is important to the successful culture of a species.

The purpose of this lab is for students to observe the development of fertilized eggs from the Japanese medaka fish.

BACKGROUND

Japanese medaka fish offer an inexpensive way to introduce students to fish reproduction. These fish can be used as an introduction to aquarium setup and maintenance before and after the fish arrive. Students are also able to work with water quality by testing the tank's water for pH and ammonia buildup. Medaka produce clear eggs that make it easy to study embryology.

The Japanese medaka, Oryzias latipes, is commonly found in Japan, Taiwan, and southeastern Asia. It is normally found in rice paddies and feeds primarily on mosquito larvae. This freshwater killifish is also known as the top minnow, geisha-girl fish, and ricefish. (See Figure Appendix G-1.)

A literal translation of the word medaka is "high-eyed." The eyes of the adult fish are close to the dorsal side of the body of the fish.

The medaka is extremely easy to maintain, as it can be kept in aquariums without a heater. It is a peaceful community fish for mixed aquariums, or it can be kept alone. Medaka breed more freely than other aquarium fish, with a single female being able to produce as many as 3,000 eggs each breeding season. Normal breeding season is June through August, but breeding can be induced almost anytime by controlling their photoperiod.

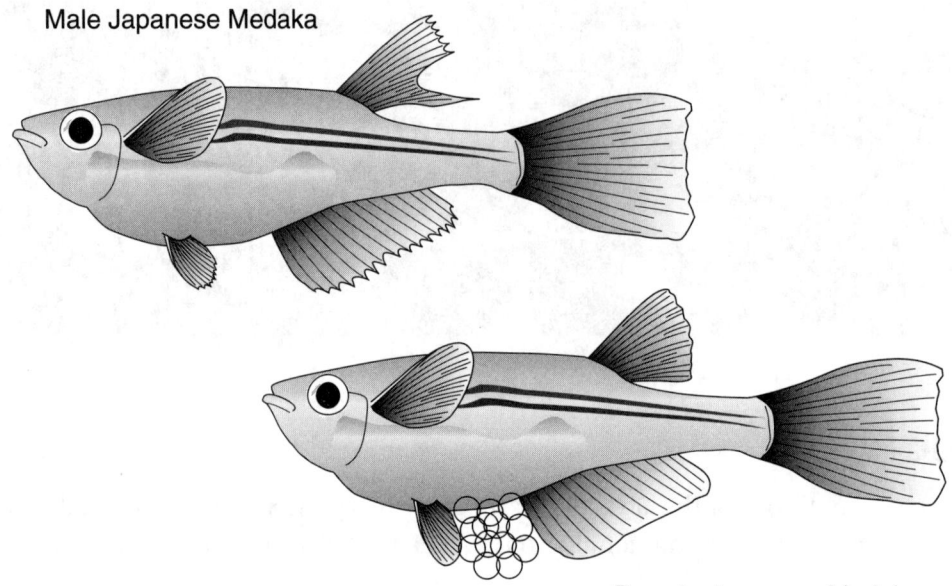

Male Japanese Medaka

Female Japanese Medaka

FIGURE APPENDIX G-1 Male and female Japanese Medaka. Note the eggs on the female.

To induce breeding, a photoperiod of at least 10 to 15 hours of light, using a simple timer on the hood lights, is required. Little, if any, ritualized courting behavior can be observed, although the most obvious sign of chasing of the female by the male prior to fertilization can be observed.

From 5 to 15 eggs are produced in a single clutch by a female with a new clutch being produced every 3 to 4 days. The eggs have a clear, transparent chorion with a number of projecting filaments that serve to attach the eggs together and to the vent of the female (see Figure Appendix G-2). A clutch (eggs laid at one time) will remain attached to a female for several hours and can be safely removed by using a soft camel-hair brush. Once removed and properly cared for, the eggs will provide an exciting study of the embryonic development of the medaka.

MATERIALS

➤ Breeding set or mixed set of Japanese medaka

➤ Dip net

➤ Aquarium

➤ Two soft camel-hair brushes

➤ 4-in culture dish

➤ Five forceps

➤ Dissecting microscope

➤ Methylene blue

➤ Dropper or pipette

➤ Petri dish

➤ Microscope slide with concave center

➤ Paper and pencil

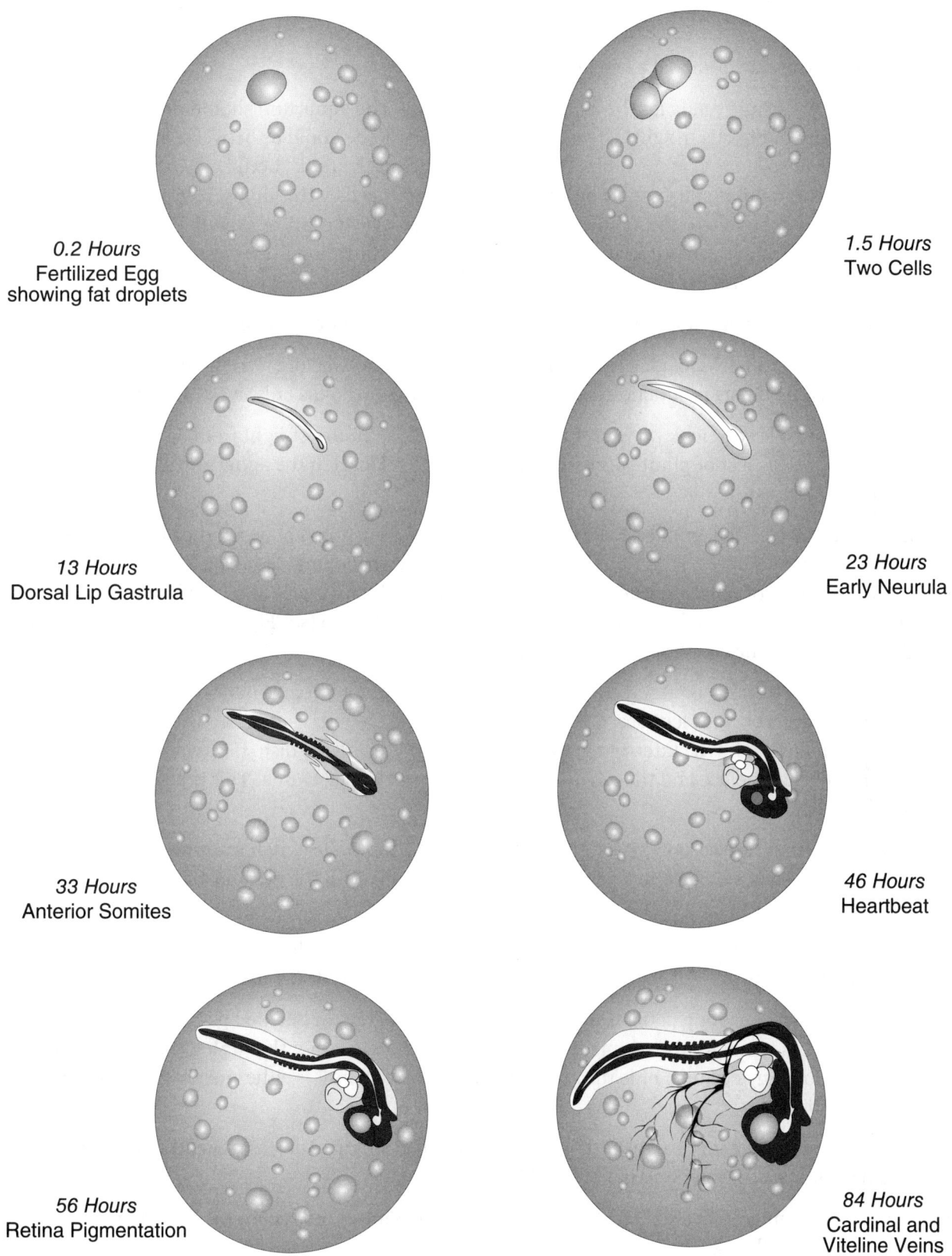

0.2 Hours
Fertilized Egg
showing fat droplets

1.5 Hours
Two Cells

13 Hours
Dorsal Lip Gastrula

23 Hours
Early Neurula

33 Hours
Anterior Somites

46 Hours
Heartbeat

56 Hours
Retina Pigmentation

84 Hours
Cardinal and
Viteline Veins

FIGURE APPENDIX G-2 Development of the fertilized eggs from the Medaka.

PROCEDURES

Specific steps in the care of medaka fish and their eggs follow:

1. Have your aquarium ready for the fish before they arrive. The aquarium should not receive more than 2 hours of direct sunlight. The temperature in the room should be between 60° and 80°F. A 5-degree drop at night is desirable. Water temperature in the tank should be kept at 75° to 80°F. With no heater, this is still possible by using the heat produced by the lamps. Live plants are beneficial, but should be kept to the middle of the tank for ease in fish capture. About 10 to 15 adult fish in a ten-gallon aquarium are recommended. Aeration and filtration are necessary. A good ratio of males to female fish is 2:3.

2. When fish are received, float the bag in the aquarium water for one to two hours.

3. Allow one or two weeks for the fish to adjust to their new home before forced breeding takes place. When breeding or egg production is not required, give the medakas only five hours of light.

4. When ready to breed, hook up your light timer for 15 to 18 hours of light from lights in the aquarium hood. Use a clear or plant-spectrum bulb. Keep the pH level in the aquarium slightly acidic for breeding.

5. Identify the males and females. Externally, males have a deep notch between the last two rays of the dorsal fin and have a larger anal fin with pronounced scalloping. Males and females are readily distinguishable. Because they are so fast in the tank, identifying males and females is difficult. Once breeding starts, egg clutches on the females are easily recognized (see Figure Appendix G-1). Also, the males chase the females around the tank.

6. Eggs will be produced about ten days after initiating the photoperiod. When you notice clear clutches of eggs on your females, prepare a 4-in culture dish with water from your aquarium. Trap a female that has a clutch of eggs and place her in the culture dish. Use two soft brushes, one to hold her in place and the other to gently brush off the egg clutch. This is a good job for two people.

7. After removing the eggs and returning the female to the tank, the clutch must now be split. You can use a microscope and a pair of fine forceps or work with the eye and forceps to break the filaments apart.

8. Prepare a mold- and bacteria-inhibiting solution with methylene blue using a concentration of 1 to 5 mg per liter of water. Fill each compartment of a divided petri dish with this solution.

9. Using a bulb dropper or pipette, transfer single eggs to the compartmented petri dish. This is important, as you will want some means of identifying each egg through embryonic development. Should any egg turn blue through this step, discard it, as it is either not fertile or contains a dead embryo.

10. The water in the rearing chamber or petri dish should be changed at least twice a week. Each time, add new aquarium water and methylene blue solution.

11. Daily, using a microscope slide with a concave center, examine and record the development witnessed in each egg.

Development of Eggs

Development of the fertilized egg requires one to three weeks depending on conditions of temperature. Because of the transparent chorion, the developmental stages are readily observed under a dissecting microscope using transmitted light and 10X to 15X magnification.

Many stages can be observed in the development of the embryo. Table Appendix G-1 describes the stages that highlight the overall development of the embryo. The descriptions in Table Appendix G-1 reflect stages of development at 77°F. Figure Appendix G-2 illustrates the development of the eggs.

Care of Newly Hatched Fry

A rearing tank of two to five gallons containing gravel, plants, and filters should be prepared during the last few days of development. Newly hatched fry should be transferred to the rearing tank with a bulb pipette. Be sure to discontinue the use of filters and bubblers when the fry are introduced into the tank; otherwise, the fry may be injured. Allow approximately 4 cm^2 of surface water for each new fish put into the tank in order to prevent overcrowding.

For the first seven to ten days, the fry should be fed Paramecia. Brine shrimp may be given as food for the next two weeks. At three weeks, dust-fine tropical fish food such as Tetramin can be fed to the fry, sparingly, twice a day.

TABLE APPENDIX G-1	**DEVELOPMENT FEATURES OF THE JAPANESE MEDAKA EGGS**	
Hours	**Features**	**Notes**
0.2	Fertilized Egg	The yolk sphere, slight yellowish color and a definite space (the perivitelline space) between the yolk and the chorionic shell.
1.5	Two Cells	Oil droplets in the yolk coalesced in the vegetal pole away from the developing embryo; cleavage of the germinal disc; recognizable cleavages occur for approximately the next 2.5 hours leading to the blastula stage.
13	Dorsal Lip Gastrula	The blastula developed to the point where invagination of the cell layers is about to occur forming the gastrula; from above, the gastrula shows a thin crease along its surface.
23	Early Neurula	The beginnings of the central nervous system evidenced by a streak of cells in the midline of the developing embryo; rapidly differentiates into the two sections of the rudimentary brain over the next 3 to 4 hours.
33	Anterior Somites	Pairs of somites (2 to 4 pairs) appear almost simultaneously near the rudimentary brain sections. The optic vesicles are apparent as enlargements on either side of the brain.
46	Heartbeat	A narrow tube ventral to the head observed to pulsate regularly; this is the early heart. Circulation and pigmentation of the blood plasma occur over the next 7 to 8 hours.
56	Retinal Pigmentation	Development of the somites and heart continues; thrashings of the tail cause the entire embryo to shift position within the shell; the retina discerned in the optic cups.
84–100	Cardinal and Vitelline Veins	Blood darkish pink; heartbeat is 130 to 150 beats per minute; frequent, sharp movements observed.
246	Hatching	Development of the major body organs including the liver, spleen, urinary and swim bladders, fins, jaw, and mouth; bright colorations characterize the various organs; before hatching, the head extends and the mouth opens.

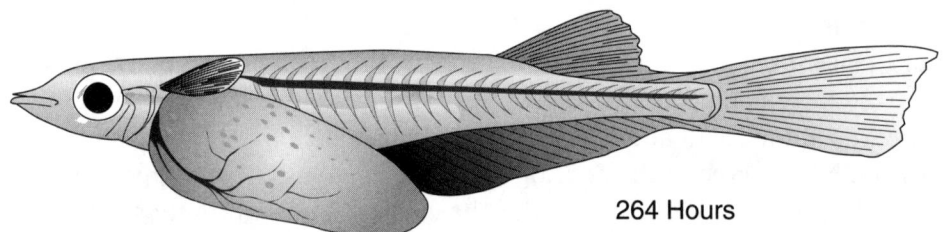

264 Hours

FIGURE APPENDIX G-3   Newly hatched Medaka fry.

Isolate newly hatched fry to avoid predation. (Figure Appendix G-3.) Six-week-old fry can be transferred to the adult tank and can be fed and treated as adults. Complete maturation occurs in two to six months. Under normal conditions, the adult medaka has a life span of four or more years.

SUPPLIERS

Breeding sets and mixed sets of Japanese medaka can be obtained from:

Carolina Biological Supply Company 2700 York Road, Burlington, NC 27215-3398
Phone: 800-334-5551
(http://www.carolina.com/)

If you do not want to care for and breed the medakas, purchase the medaka egg set. These can be maintained for the observation of embryo development.